Workshop on Oxide Materials 2014: Novel Multifunctional Properties

Journal of Physics: Conference Series
Volume 614

Cali, Colombia
15 - 19 September 2014

Editors:

Maria Elena Gomez
Wilson Lopera

ISBN: 978-1-5108-0754-9
ISSN: 1742-6588

Printed from e-media with permission by:

Curran Associates, Inc.
57 Morehouse Lane
Red Hook, NY 12571

Some format issues inherent in the e-media version may also appear in this print version.

Printed by Curran Associates, Inc. (2015)

For permission requests, please contact the Institute of Physics
at the address below.

Institute of Physics
Dirac House, Temple Back
Bristol BS1 6BE UK

Phone: 44 1 17 929 7481
Fax: 44 1 17 920 0979

techtracking@iop.org

Additional copies of this publication are available from:

Curran Associates, Inc.
57 Morehouse Lane
Red Hook, NY 12571 USA
Phone: 845-758-0400
Fax: 845-758-2634
Email: curran@proceedings.com
Web: www.proceedings.com

Table of contents

Volume 614

Workshop on Oxide Materials 2014: Novel Multifunctional Properties
15–19 September 2014, Cali, Colombia

Accepted papers received: 10 April 2015
Published online: 3 July 2015

Preface

011001

Workshop on Oxide Materials 2014: Novel Multifunctional Properties OPEN ACCESS
M E Gómes and W Lopera

011002

Peer review statement OPEN ACCESS

Papers

Synthesis and Characterization of Oxides

012001

Effect of fluoride and water content on the growth of TiO2 nanotubes synthesized via ethylene glycol with voltage changes during anodizing process OPEN ACCESS Heiddy P Quiroz, Francisco Quintero, Pedro J Arias, A Dussan and Hugo R Zea pg. 1

012002

Synthesis, structural characterization, electric and magnetic behaviour of a Sr_2DyNbO_6 double Perovskite OPEN ACCESS P C Plazas Hurtado, D A Landínez Téllez, J A Cardona Vásquez, C A Parra Vargas and J Roa-Rojas pg. 8

012003

Production and structural and magnetic characterization of a $Bi_{1-x}Y_xFeO_3$($x = 0$, 0.25 and 0.30) system OPEN ACCESS J A Mejía Gómez, G I Supelano García, C A Palacio and C A Parra Vargas pg. 13

Magnetic Fluids

012004

Magnetic Study of Fe_3O_4 Particles for Implementation as Magnetic Fluids OPEN ACCESS H A Fonseca, E Gonzalez and C A Parra Vargas pg. 18

Quantum Magnetism

012005

Quantum magnetism in spin-3/2 chains OPEN ACCESS J J Hernández-Sarria, A Argüelles and K Rodríguez pg. 22

012006

Spin polarization in non-magnetic nanostructures OPEN ACCESS J A Zúñiga, S T Pérez Merchancano and L E Bolívar Marínez pg. 28

Catalytic Activity

012007

Preparation of Fe_3O_4 Nanoparticles and Removal of Methylene Blue through Adsorption OPEN ACCESS J S Trujillo Hernandez, A Aragón Muriel, J A Tabares, G A Pérez Alcázar and A Bolaños pg. 35

012008

Effect of annealing treatment on the photocatalytic activity of TiO_2 thin films deposited by dc reactive magnetron sputtering OPEN ACCESS L M Franco Arias, A Arias Duran, D Cardona, E Camps, M E Gómez and G Zambrano pg. 39

Hybrid Multiferroic

012009

Detailed Study of Structural and Ferroelectric Properties in $BaTiO_3/La_{2/3}Ca_{1/3}MnO_3$ Bilayers Deposited by Pulsed-Laser Deposition OPEN ACCESS J E Ordoñez, M E Gomez, W Lopera and P Prieto pg. 45

Journal of Physics: Conference Series **614** (2015) 011001 doi:10.1088/1742-6596/614/1/011001

The 2014 Workshop on Oxide Materials: Novel Multifunctional Properties was held in Cali, Colombia, from September 15 to September 19 on the campus of *Universidad del Valle*. It was a great privilege to have had this workshop in Cali after the first workshop on oxide materials commemorating the first centennial of the discovery of the superconductivity in 2011. The meeting gathered an audience of 80 participants, 10 invited speakers with two or three plenary talks each, 20 short oral contributions, two poster sessions with 20 presentations each. Plenary sessions were offered by the following invited speakers:

Ivan K. Schuller, Director of the Center for Advanced Nanoscience, University of California San Diego, USA.

- Simple oxides
- Strongly Correlated Degrees of Freedom in Complex Oxides and the Discovery of New Physics

Axel Hoffmann, Argonne National Lab., Illinois, USA.

- Properties of Magnetic Heterostructures, Spin-Transport in Novel Geometries, and Biomedical Applications of Magnetism

Ricardo Ibarra, Director *Instituto de Nanociencia de Aragón*, University of Zaragoza, Spain

- Functional Magnetic Oxide Nanostructures
- Anomalous Hall, Nerst and Thermo-Spin Effect in Epitaxial Magnetite
- Magnetic Oxide-Based Nanoparticles for Hyperthermia-Induced Cell Death

Juan C. Nino, Department of Materials Science and Engineering, University of Florida at Gainesville, USA

- The Unique Role of Oxides in the Advancement of Energy Technologies
- Crystallography of Oxides for Electronic Applications
- Recent Developments in Ionic and Protonic Conduction in Ceramics

Gabriel Ramirez, Department of Physics, University of California at San Diego, USA

- Metal-to-Insulator Transition in VO_2 Thin Films: Structural Evolution and Long-Range Correlation
- Dynamics and Magnetics in High-Quality V_2O_3 Thin Films

Sara Aldabe Bilmes, *Departamento de Química Inorgánica, Analítica y Química Física,* Universidad de Buenos Aires, Argentina

- Chemistry of Precursors, Nucleation and Growth
- Particles in Suspension: Stability, Surface, Surface Functionalization
- Applications: Catalysis, Encapsulation of Biomolecules, Cells and Microorganisms, Sensors

Juan Muñoz Saldaña, Senior Researcher, *Centro de Investigación y de Estudios Avanzados del Instituto Politécnico Nacional (CINVESTAV),* Queretaro, Mexíco

- Reconstruction of Domains in Ferroelectric Perovskite Ceramics by Piezoresponse Microscopy

- Ferroelasticity and Redirect Domains in Multifunctional Ceramics (Zirconia and Perovskite-Based Ceramic)
- Lead-Free Ferroelectric Bismuth Basis

Anderson Dussan Cuenca, Department of Physics, Universidad Nacional de Colombia
- Solar Cells
- Magnetic Diluted Semiconducting Materials

Karem Rodríguez, Department of Physics, *Universidad de Valle,* Colombia
- Quantum Magnetism
- Strongly Correlated Systems

Pedro A. Prieto, Department of Physics, *Universidad del Valle, Colombia*
- Novel Oxides from Superconductor, Magnetic and Multiferroic materials

This proceedings volume contains papers reported at the conference. The Proceedings of the 2014 Workshop on Oxide Materials: Novel Multifunctional Properties were edited by Maria Elena Gomez and Wilson Lopera with the assistance of Carlos William Sanchez and Albert Ortiz as copy editor.

We are grateful for the financial support from COLCIENCIAS through research project COLCIENCIAS-UNIVALLE contract 002/2013; Universidad de Valle through Professor Ivan Ramos, Rector; the Faculty of Science with Professor Jaime Cantera, Dean; the Center of Excellence on Novel Materials with Professor Pedro Prieto, Director; ICETEX, and INTECO Ltda.

Maria Elena Gómez, Editor
Wilson Lopera, Editor

Participants of Workshop on Oxide Materials: Novel Multifunctional Properties

Peer review statement

All papers published in this volume of *Journal of Physics: Conference Series* have been peer reviewed through processes administered by the proceedings Editors. Reviews were conducted by expert referees to the professional and scientific standards expected of a proceedings journal published by IOP Publishing.

Effect of fluoride and water content on the growth of TiO_2 nanotubes synthesized via ethylene glycol with voltage changes during anodizing process

Heiddy P. Quiroz[1], Francisco Quintero[2], Pedro J. Arias[2], A. Dussan[1], Hugo R. Zea[2]

[1]Departamento de Física, Grupo de Materiales Nanoestructurados y sus Aplicaciones, Universidad Nacional de Colombia, Colombia.

[2]Grupo de Investigación en Materiales, Catálisis y Medio Ambiente, Universidad Nacional de Colombia, Facultad de Ingeniería, Departamento Ingeniería Química y Ambiental, Carrera 30 No 45-03, Bogotá, DC, Colombia

E-mail: adussanc@unal.edu.co

Abstract. In this work, titanium foils were anodized in ethylene glycol solutions containing different amounts of water and fluoride to determine their effects on the top morphology and crystalline structure of the formed titania nanostructures. Anodizing was performed for 2 h by using titanium foils as both anode and cathode applying a squared-pulse voltage profile composed of one step at 80 V for 3 min followed by another at 20 V for 5 min; constant voltage conditions were also used to study the nanostructure formation on the surface. We found the formation of nanostructured titania on the surface of the anodized foil when small amounts of water and fluoride are present in the anodizing solution. The top of these nanostructures is irregular when no water is added, but is expected to change with different amounts of water and fluoride in the ranges of 1 - 9 % and 0.05 - 0.5 %, respectively. Synthesis parameters also change nanotube morphology. The morphology and structure properties of the samples were studied by scanning electron microscopy (SEM) and X-ray powder diffraction (XRD). Formation of TiO_2 nanotubes by anodization method are strongly correlated to conditions like fluoride concentration and applied voltages. Tube length varying between 2 and 7 μm, exhibiting different diameters and wall thicknesses were obtained. When an alternate voltage was applied, the wall of the nanotubes presented evenly spaced rings while nanotubes with smooth wall form were observed when constant voltage was applied. Reflection peaks corresponding to Brookite, Anatase, and Rutile of TiO_2 phases were observed from XRD measurements. A correlation between the effects of synthesis parameters on nanotube formation and morphological properties is presented. TiO_2 nanotubes prepared by electrochemical anodization have excellent performance in various applications such as photocatalysts, solar cells, gas sensors, and biomedical applications.

1. Introduction

The main achievements of technology based on foils or semiconductor thin films and oxide materials have contributed to the development of optical, microelectronic, and nanoscale devices. This has generated important interest in the study of optical, structural, mechanical, and electrical properties for several applications [1]. One-dimensional nanostructures, like nanotubes, nanowires and nanorods began with the discovery and development of carbon nanotubes [2] and their high impact upon nanotechnology systems due to their molecular geometry and interesting properties such as high electron mobility, the effects of quantum confinement, and high specific surface area, among others [3-5].

Titania or titanium dioxide (TiO_2), as part of oxide materials, is used in different applications including photocatalysis [6], dye sensitized solar cells (DSSC) [7], gas sensors [8], paints [9], biomedical devices [10], and the food industry [11]. Compared to other methods such as chemical vapor deposition (CVD) [12], sol–gel synthesis [13], hydrothermal [14] and magnetron sputtering deposition [15], the electro-chemical anodization method is regarded as one of the simplest techniques to obtain TiO_2 nanotubes with large internal surface area without a concomitant decrease in geometric and structural order [16].

Anodization parameters play an important role on the titania nanotube formation with characteristics defined in their form, length, pore or tube diameter, thickness and pore-pore distance. Titanium dioxide nanotubes by electrochemical anodization method were reported using a titanium foil in acid electrolyte HF [17]. Other studies have reported that the crystalline structure of the titania nanotubes is characterized by a mix of anatase and rutile phases that changes with the annealing process [18].

This work presents a detailed study of the dependence of anodization parameters on TiO_2 nanotubes by electrochemical anodization method, using ethylene glycol with voltage changes during the anodizing process. Structural and morphological properties were studied via X-ray diffraction (XRD) and scanning electron microscopy (SEM). It was observed that tube length varying between 2 and 7 μm, exhibited different diameters and wall thicknesses formations. When alternate voltage was applied, the outer wall of the nanotubes presented evenly spaced rings, while it was smooth when constant voltage was applied.

2. Experimental procedure

Prior to the anodization process, the Ti foils used in this study were cleaned with methanol at room temperature and air dried after that. The titanium foils used in this work were a 99.97% pure. The TiO_2 nanotubes were prepared from Ti substrates 2 x 4 cm in size with 50 μm thickness. An ethylene glycol solution, deionized water, and ammonium fluoride (NH_4F) were used in the mixture. All samples were fabricated at 300 K.

During the first stage of fabrication, NH_4F was dissolved in deionized water and mixed in 126 ml of ethylene glycol while being stirred at 120 rpm until obtaining a clear and uniform solution; subsequently, two Ti foils that work as the anode and cathode with 2 cm separation from each other were introduced into the solution while being stirred. Nanotubes were obtained from a solution containing between 0.05 - 0.5 w% of NH_4F and 1 - 9% of H_2O in ethylene glycol for 2 h. The anodizing process was performed by applying a squared-pulse voltage profile composed of one step at 80 V for 3 min, followed by another one at 20 V for 5 min. Additionally, constant voltage conditions were also used to study the nanostructure formation on the surface.

The samples were characterized through XRD measurements by using an Empyrean diffractometer (PANalytical) equipped with a Co-Kα source: 1.7890 Å, 40 kV and 40 mA with a Bragg-Brentano configuration and Pixcel^{3D} 2X2 detector. X'Pert High-Score Plus software was used to compare the samples through Rietveld refined in the range of $10° \leq 2\theta \leq 140°$ with angular steps of 0.0001°. The morphological characterization of the samples, through SEM measurements, was performed by using a microscope (VEGA3 SB) with a tungsten filament, an accelerating voltage of 4.89 kV in low vacuum conditions (10^{-3} mbar), and a XFlash Detector 410M.

3. Results and discussion

Figure 1a shows XRD patterns of the TiO_2 samples changing from 0.05 to 0.1% ammonium fluoride concentration and maintaining other parameters constant: 9% H_2O, 126 ml ethylene glycol, T= 300K, squared-pulse voltage profile 80 – 20 V. In this figure the XRD pattern corresponding to the samples after the annealing process to T= 723K during 1 h. It is well known that an amorphous nature in all samples is presented when the samples are prepared by anodization method and, subsequently, their crystal structure changes with a high temperature anneal around 700 K [19]. Higher temperature anneals improve crystallinity, showing a mix of anatase and rutile phases on nanotube walls. However, in this case 9% of H_2O into ethylene glycol solution, TiO_2 nanotubes formation on surface weren't observed. (Figure 1b). Figure 1b and 1c show SEM images of the TiO_2 samples with 0.1 and 0.05% NH_4F concentration. It can be seen that the surface, after annealing at 723 K, is characterized by a uniform morphology and the absence of nanostructures with nanotube shape or other structures.

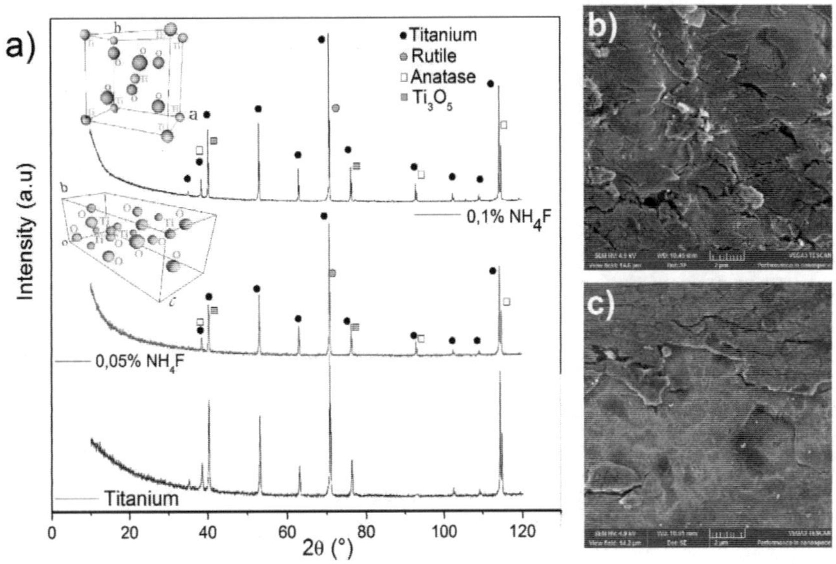

Figure 1. a) XRD of the TiO_2 obtained with 9% of H_2O in ethylene glycol solution changing ammonium fluoride concentration: 0.10 % (black line) and 0.05 % (red line), Ti foil pattern (blue line); SEM images with b) 0.10 % and c) 0.05 % NH_4F. Inset shows atomic positions obtained for anatase and rutile phases.

From Figure 1, it was observed that crystalline structures are characterized mostly by the presence of Ti and rutile phases in relation to the anatase phase in the compound. Peaks located at 42.07° and 78.35° (Figure 1a) were associated to the Ti_3O_5 phase, possibly formed during the anodization process due to high water concentration in the ethylene glycol solution. **The** Ti_3O_5 phase is a transition metal oxide with an octahedral formation, where Ti is the transition metal ion with 3d orbitals and known to undergo phase transitions of crystal structures, electron transport, or magnetic properties. This transition is of the first order accompanied by a rapid decrease of resistivity and a rapid increase of the c axis and magnetic susceptibility, caused by heating to a transition temperature of about 460 K on a sample containing considerable amounts of elements other than titanium and oxygen [20]. Rutile and anatase phases were identified from PDF files 01-088-1174 and 00-021-1272, respectively, and compared to peak positions obtained by XRD measurements (Figure 1); that is, 70.87° for the rutile phase and 114.82° for the anatase phase.

The inset in Figure 1a shows the crystal structure of the anatase and rutile phases simulated from the crystalline parameters (obtained from XRD) and reported in Tables 1 and Tables 2, respectively.

Table 1. Crystalline parameters of anatase phase.

Crystalline parameters					Atomic Positions			
a (Å)	**b (Å)**	**c (Å)**	**Space Group**	**N° Space Group**	**Element**	**X**	**Y**	**Z**
3,77	3,77	9,48	$I4_1$/amd	141	Ti	0,25	0,375	0
α (°)	**β (°)**	**ϒ (°)**	**Volume (Å^3)**	**Crystalline System**	O	0,25	0,167	0
90	90	90	135,25	Tetragonal				

Table 2. Crystalline parameters of rutile phase.

Crystalline parameters					Atomic Positions			
a (Å)	**b (Å)**	**c (Å)**	**Space Group**	**N° Space Group**	**Element**	**X**	**Y**	**Z**
4,53	4,53	2,92	$I4_2$/amd	136	Ti	0	0	0
α (°)	**β (°)**	**ϒ (°)**	**Volume (Å^3)**	**Crystalline System**	O	0,3039	0,3039	0
90	90	90	56,94	Tetragonal				

Figure 2 shows X-ray spectra of the TiO_2 samples, changing H_2O concentration in the ethylene glycol solution by 9% (blue line) and 1% (black line). The other parameters were maintained constant: ammonium fluoride concentration of 0.05%, squared-pulse voltage profile composed of one step at 80 V for 3min, followed by another one at 20 V for 5 min; anodization time of 2 h and environment temperature during the anodization process. Anatase and rutile phases were observed in all samples after annealing at T = 723 K.

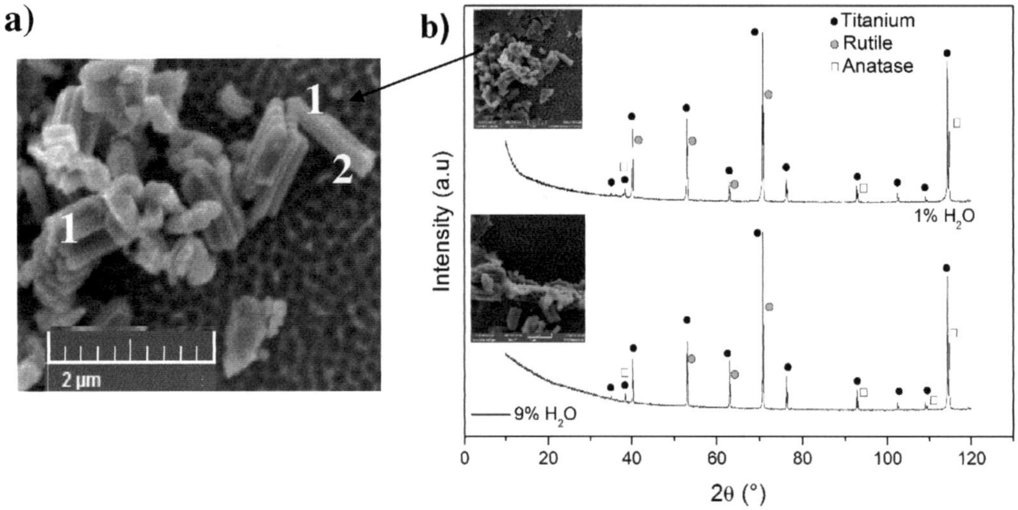

Figure 2. a) Magnification of SEM micrograph for the sample with 1% H_2O (where 1- is tubes closed at the bottom and 2- is tubes open at the top.) and **b)** X-ray diffractogram of the TiO_2 obtained with 1% (black line) and 9% (blue line) H_2O in ethylene glycol solution. In the inset, SEM images are showed for each one.

For main peak 2θ = 70, 69° (see Figure 2) is possible to observe Ti and anatase phases contribution for all samples; while, Ti_3O_5 phase was not observed due to low amount of oxygen in the solution. Other studies has been reported with a similar behavior [20].

On the other hand, the inset in Figure 2b shows SEM images for TiO_2 samples with 0.15 and 0.05% ammonium fluoride concentration, prepared at an alternate voltage of 80 – 20 V. Figure 2a was shows the surface exhibits the tubes open on the top and closed at the bottom, nearly parallel and closed down. At the bottom on the surface of Ti foil, traces of the tubes were observed when they were detached from the solid TiO_2 layer.

Figure 3 shows the XRD pattern of the TiO_2 sample when subjected to the annealing process varying temperature between 293 and 723 K. A sample was prepared by using NH_4F concentration varying from 0.05% to 0.25% (2% H_2O into solution) and constant voltage of 80V for 45 min. From the annealing process a decrease in the amorphous fraction of the material was observed, together with a change in the crystalline structure. An anatase phase was obtained from a TiO_2 sample with a high presence in the material at T = 723 K.

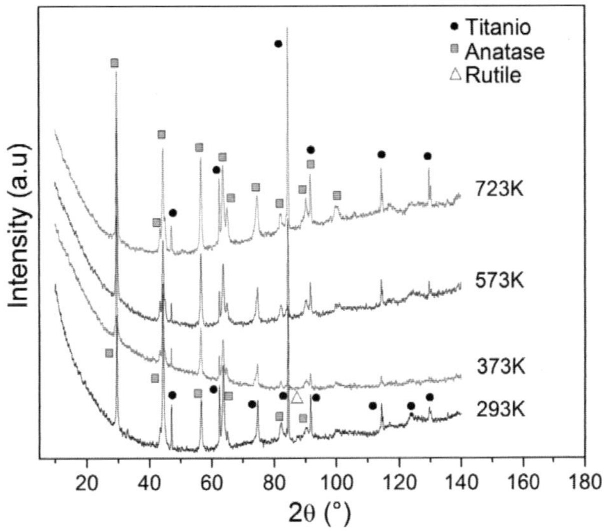

Figure 3. XRD pattern for theTiO$_2$ sample prepared with 2% H_2O, 0.25% NH_4F, and constant voltage 80 V.

Both, alternate and constant voltage conditions were used to study the topographic and morphological properties of nanotube formation on the surface; it has been reported that sidewall inhomogeneity is associated to current fluctuations [16], but studies on voltage change on nanotube formation have yet to be reported. It was observed that, when alternate voltage was used for nanotube preparation, these were small and thick-walled (Figure 2a); while, the application of a constant voltage allowed for nanotube formation with a length of 2.97 ± 0.19 μm with thin walls (Figure 4e).

Figure 4a-d) shows SEM micrographs of the TiO_2 sample subjected to annealing process and shown in Figure 3.

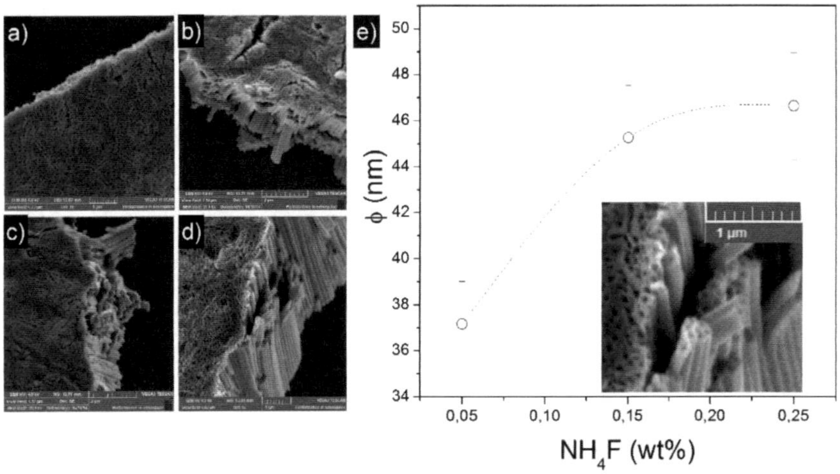

Figure 4.a-d) SEM micrographs of the TiO_2 sample varying annealing temperature 293, 373, 573, and 723 K; e) variation of inner diameter of titania nanotubes ($\phi \pm 4.30$ nm).

In these images (Figure 4), surface morphology is not altered by annealing times, as observed in the crystal structure (Figure 3). The inner diameter size for nanotubes was obtained, varying between 37 and 46 nm when NH_4F concentration varied from 0.05 to 0.25% (Figure 4e). This indicates that the size of the nanotube diameter is greatly dependent on anodization parameters and, in this case, on the NH_4F concentration for a constant voltage. From SEM images, values close 2 μm in length and 29.56 nm in thickness of barrier layer were obtained with the nanotube arrays grown vertically on the Ti foil.

4. Conclusions

Titanium dioxide nanotube arrays were successfully obtained by electrochemical anodization method, varying both voltage and NH_4F concentration from 0.05 to 0.25% and the amount of water into the ethylene glycol solution. From the XRD pattern, the peaks corresponding to anatase and rutile phases were identified for a temperature range between 293 and 723 K. Irregular nanostructures were observed in different amounts of water and fluoride solutions in the ranges of 1 to 9 % and 0.05 to 0.25 %, respectively. From SEM measurements, the diameter of the nanotubes was obtained varying between 37 and 46 nm when NH_4F concentration varied from 0.05 to 0.25%. Values around 2.98 ± 0.19 μm length and 29.56 ± 1.93 nm in diameter were obtained with constant voltage.

5. Acknowledgments

This work was supported by Universidad Nacional de Colombia – DIB, Project Cod. Quipú No. 201010020493 and 201010020453.

References

[1] José M. Albella 2003 Láminas delgadas y recubrimientos: Preparación, propiedades y aplicaciones (España, Editorial CSIC).
[2] S. Iijima 1991 *Nat.* **354** 6-58.
[3] C.M. Lieber 1998 *Solid State Commun.* **107** 607-616.
[4] C. N. R. Rao, A. Miller, A. K. Cheetham 2006 *The Chemistry of Nanomaterials: Synthesis, Properties and Applications* (Oxford University, Wiley-VCH) Vol. 1.
[5] C. Weisbuch, B. Vinter 1991 *Quantum Semiconductor Structures: Fundamentals and Applications* (San Diego, California, Elsevier).
[6] Jian-Ying Huang, Ke-Qin Zhang, Yue-Kun Lai 2013 Int. J. Photoenergy **2013** 1-19.
[7] Junfeng Yana, Feng Zhou 2011 J. Mater. Chem. **21** 9406-9418.

[8] Yongxiang Li, Xiaofeng Yu, Qunbao Yang 2009 J. Sens. **2009** 1-19.
[9] Ricardo A.R. Monteiro, Filipe V.S. Lopes, Adrián M.T. Silva, Joana Ângelo, Gabriela V. Silva, Adélio M. Mendes, Rui A.R. Boaventura, Vítor J.P. Vilar 2014 Appl. Catal. B **147** 988–999.
[10] Craig A. Grimes and Gopal K. Mor 2009 *TiO_2 Nanotube Arrays Synthesis, Properties, and Applications* (New York, Springer).
[11] Gizem Karaca, Yücel Tasdemir 2014 Sci Total Environ **488–489** 356–361.
[12] Akihiko Ito, Takahiro Sato, Takashi Goto 2014 Thin Solid Films **551** 37–41.
[13] Y.F. You, C.H. Xu, S.S. Xu, S. Cao, J.P. Wang, Y.B. Huanga, S.Q. Shi 2014 Ceramic. Int. **40** 8659–8666.
[14] Yanyan Xu, Maocui Zhang, Miao Zhang, Jianguo Lv, Xishun Jiang, Gang He, XuePing Song, Zhaoqi Sun 2014 Appl. Surf. Sci. **315** 299–306.
[15] A. Lin, W. C. Lu, C. Y. Wu, K. S. Chang 2014 Ceramic. Int. **40** 15523–15529.
[16] Poulomi Roy, Steffen Berger, Patrik Schmuki 2011 Angew. Chem. Int. Ed. **50** 2904–2939.
[17] Guohua Liu, Kaiying Wang, Nils Hoivik, Henrik Jakobsen 2012 Sol. Energy Mater. Sol. Cells **98** 24–38.
[18] Rajini P.Antony, TomMathews, ArupDasgupta, S.Dash, A.K.Tyagi, BaldevRaj 2011 J. Solid State Chem. **184** 624–632.
[19] Jongyun Moon, Marianna Kemell, Byungki Park, Arho Suominen, Ermei Mäkilä, Risto Punkkinen, Hannu-Pekka Hedman, Hong Kim, Lippo V Lassila and Aulis Tuominen 2013 Color. Technol. **130** 1–7.
[20] Masashige Onoda 1998 J. Solid State Chem. **136** 67–73.

Synthesis, structural characterization, electric and magnetic behaviour of a Sr_2DyNbO_6 double Perovskite

P.C. Plazas Hurtado[1], D.A. Landínez Téllez[1], J.A. Cardona Vásquez[1], C. A. Parra Vargas[2], J. Roa-Rojas[1]

[1] Grupo de Física de Nuevos Materiales, Departamento de Física, Universidad Nacional de Colombia, AA 5997, Bogotá DC, Colombia
[2] Grupo Física de Materiales, Escuela de Física, Universidad Pedagógica y Tecnológica de Colombia, Tunja, Colombia

E-mail: pcplazash@unal.edu.co

Abstract. Many multifunctional properties have been the focus of our interest in Perovskite research. Herein, we report synthesis and characterization of a new Sr_2DyNbO_6 Perovskite-like material. Samples were produced via standard solid-state reaction method. Rietveld refinement of X-ray diffraction experimental data through the GSAS code reveals that this material crystallizes in a monoclinic complex Perovskite (space group $P2_1/n$, #14) with lattice parameters $a=5.730(3)$ Å, $b=5.905(1)$ Å and $c=8.221(0)$ Å. Scanning electron microscopy images reveal that samples evidence a surface with strongly diffuse granular structure. From curves of magnetization as a function of temperature, we determined the paramagnetic behaviour of this complex Perovskite in the temperature regime between 50 and *320 K*. Curie fitting allows obtaining an effective magnetic moment of $10.28\ \mu_B$. Polarization curves as a function of the applied electric field show a hysteretic behavior of typical dielectric loss with relative dielectric constant of 264.28 at room temperature.

Keywords: Complex Perovskite, structure, magnetic behaviour

Introduction

In recent years, much interest has emerged in electronic oxide ceramic oxides with ABO_3 Perovskite structure due to their chemical versatility to evaluate different combinations of transition metal cation in the B site and obtain a variety of magnetic and electric properties [1]. Ideally, a simple ABO_3 Perovskite possesses a tri-dimensional cubic network of BO_6 octahedra with the cation A into the void formed by eight octahedra with AO_{12} coordination [2]. If the A and B sites contain a mixture of cations, distorted complex Perovskite structures with lower symmetries are obtained. Distortions are mainly related to octahedral tilting due to small cation driving a rotating of octahedra with respect to the crystallographic axes. Another kind of distortion is the cation displacement caused through first- and second-order Jahn-Teller effects. Specially, the 1:1 B-site based class with chemical formula $A_2BB'O_6$ (or double Perovskites) is the most frequently encountered complex Perovskite [3]. From the distortions and composition of the double Perovskites, new properties and applications are found in ferroelectric devices with large dielectric constants and frequency dispersion [4], multiferroicity [5], solar energy conversion [6], catalysis [7], solid-oxide fuel cells [8], and microwave resonator [9]. Previous studies have shown that under certain conditions, the double Perovskite Sr_2RESbO_6 (RE=Dy,

Ho, Gd) crystallizes in geometrically frustrating fcc lattices [10]. In this study, the authors report that the RE^{3+} and Sb^{5+} cations adopt an ordered array in the double-Perovskite structure with alternate octahedra. For RE=Dy, tfrustrated magnetic ordering has been observed. In order to study the effect of the substitution of Nb^{5+} in the crystallographic site of the Sb^{5+} on the crystal structure and magnetism, we synthesize the non-studied Sr_2DyNbO_6 double Perovskite. On the other hand, due to the paramagnetic insulator behaviour reported at room temperature for the Sr_2RESbO_6 [10], the electric polarizability of this material is particularly interesting.

Experimental

Samples were synthesized by means of the standard solid-state reaction recipe. The precursor powders $SrCO_3$, Dy_2O_3, and Nb_2O_5 (Aldrich 99.9%) were stoichiometrically mixed according to the chemical formula Sr_2DyNbO_6. The mixture was annealed at 900 °C for 12 h. The samples were then reground, re-pelletized, and sintered at 1350 °C for 36 h. X-ray diffraction (XRD) experiments were performed by means of a PW1710 diffractometer with $\lambda_{CuK\alpha}$=1.54064Å. Rietveld refinement of the diffraction pattern was made by the GSAS code [11]. Field cooling measurement of magnetization as a function of temperature was carried out by using VersaLab Quantum Design equipment. Morphological studies were performed via scanning electron microscopy (SEM) experiments through the utilization of a FEI Quanta 200 microscope. Polarization curves as a function of the applied electric field were obtained through a Radiant ferroelectric tester on the application of voltages up to 1000 V.

Results and discussion

Figure 1 shows the XRD pattern for the Sr_2DyNbO_6 double Perovskite at room temperature. It is clear from Figure 1 that the sample has a single crystallographic phase. From the Rietveld [11] it was determined that Sr_2DyNbO_6 crystallizes in a monoclinic structure, space group $P2_1/n$ (#14), which denotes a primitive cell with two-order screw axis (180°) rotation, followed by a translation of half lattice parameter in the [010] direction, and a glide plane reflection perpendicular to the [010] direction. The structure obtained is shown in Figure 2. The lattice parameters obtained were a = 5.7303 Å, b = 5.9051 Å, and c = 8.2210 Å. The presence of the (211), (3-10), and (133) peaks in the diffraction pattern is an evidence of the possible cationic order of Dy^{3+} and Nb^{5+} in B and B' sites, respectively, of the $A_2BB'O_6$ double Perovskite.

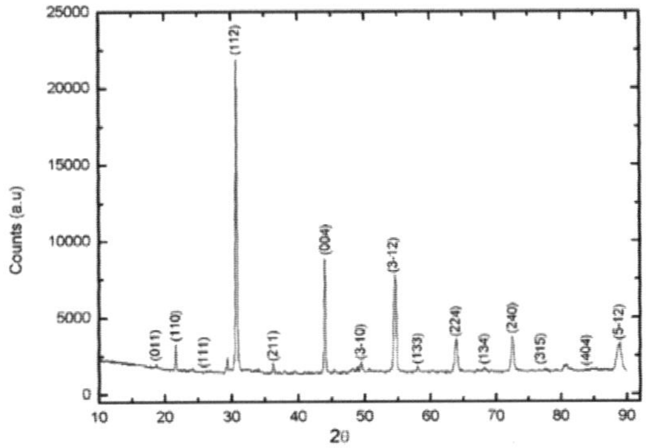

Figure 1. XRD pattern for the Sr_2DyNbO_6 complex Perovskite

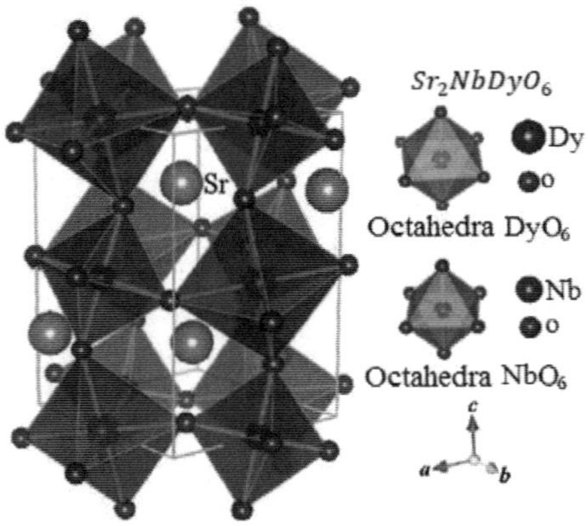

Figure 2. Crystalline structure of the Sr_2DyNbO_6 double Perovskite

Figure 3 exemplifies the surface morphology of the Sr_2DyNbO_6 sample. The SEM image shows homogeneous granularity, conformed by strongly diffused grains whose size varies from submicrometric up to micrometric scale (~ 3 μm).

Figure 3. SEM image of the surface structure for the Sr_2DyNbO_6 sample

Measurements of magnetic susceptibility as a function of temperature on the application of H = 10 Oe reveals the paramagnetic behaviour of this material in the temperature regime from 50 to 300 K (results are shown in Figure 4).

From the fit of the susceptibility curve with the Curie behaviour given by $\chi=C_c/T$, we obtain a Curie constant $C_c=13.2423$ *emu·K/mol*, which corresponds to an effective magnetic moment of $\mu_{eff}=10.28$ μ_B. This value agrees with the expected value of $\mu_{eff}=10.63$ μ_B predicted by Hund's rules for the isolated Dy^{3+} cation [12]. Figure 5 shows the hysteric behaviour of the electric polarization on the application of electric fields up to *3.0 kV/cm*. The characteristic curve reveals strong dielectric loss with a saturation polarization of *0.065* $\mu c/cm^2$, remnant polarization of *0.039* $\mu c/cm^2$ and coercive field of *1.48 kV/cm*. From the estimated values of P_S, we determined the capacitance of the Sr_2DyNbO_6 material through the equation $P_s = (C/A)V$, where A represents the area of the sample, V is the applied voltage, and C is the capacitance. Then, we obtained C/A by taking into account the range of high applied field of Figure 5 curve. The dielectric constant is calculated by the expression $\varepsilon = (C/A)(d/\varepsilon_0)$, where d represents the distance between the capacitor plates and ε_0 is the dielectric constant in vacuum [13]. The value obtained of the relative dielectric constant at room temperature for the SZMO complex Perovskite is $\varepsilon=264.28$, which is five times the value reported for $BaZrO_3$ [14].

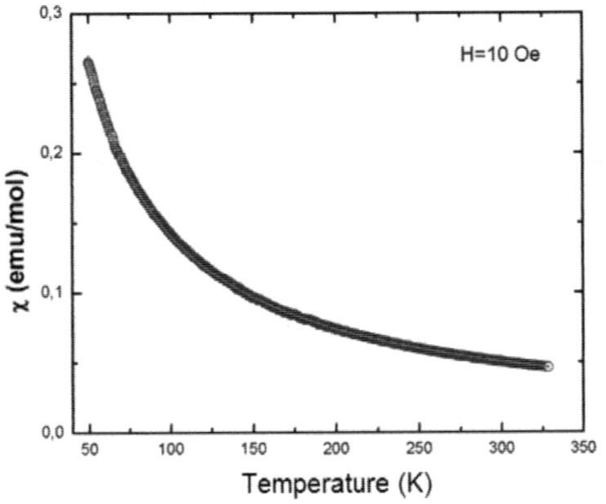

Figure 4. Curve of susceptibility as a function of temperature for the Sr_2DyNbO_6 perovskite

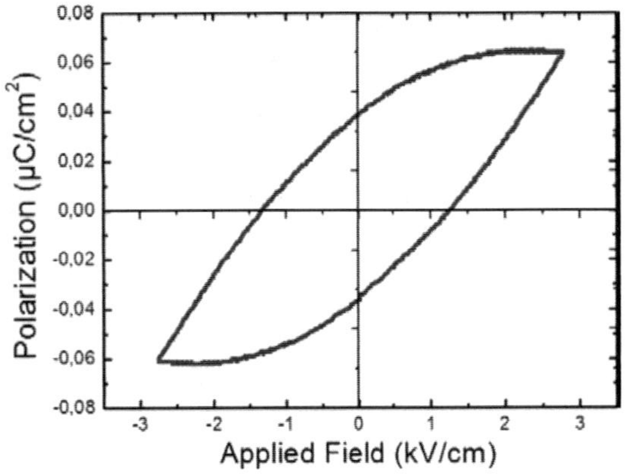

Figure 5. Hysteretic feature of the electric polarization for the Sr_2DyNbO_6 Perovskite

Conclusions

Synthesis and structural characterization of the new Sr_2DyNbO_6 Perovskite-like material were performed. Rietveld analysis of experimental diffraction pattern reveals that this material crystallizes in a monoclinic complex Perovskite, which belongs to the space group $P2_1/n$ (#14). The SEM images reveal the strongly diffuse granular surface morphology. Measurements of magnetic susceptibility show the paramagnetic behaviour of the material in the temperature interval between 50 and 300 K. The experimental effective magnetic moment agrees with the theoretical value expected for the isolated Dy^{3+} ion. The saturation of polarization in the dielectric hysteresis curve permitted determining the relative dielectric constant at room temperature $\varepsilon=264.28$, which is an appropriate value for microwave applications.

Acknowledgements

This work was partially supported by the Research Division (DIB) at Universidad Nacional de Colombia and *"El Patrimonio Autónomo Fondo Nacional de Financiamiento para la Ciencia, la Tecnología y la Innovación Francisco, José de Caldas"* – COLCIENCIAS, contract RC-No. 0850-2012.

References

[1] C. Franchini, 2014, J. Phys. Condens. Matter **26**, 253202.
[2] L.C. Moreno, J.S. Valencia, D.A. Landínez Téllez, J. Arbey Rodríguez M., M.L. Martínez, J. Roa-Rojas, F. Fajardo, 2008, J. Magn. Magn. Mater. **320**, e19-e21.
[3] C. J. Howard, B. J. Kennedy, P. M. Woodward, 2003, Acta Cryst. **B59**, 463-471.
[4] E. J. Juarez-Perez, R. S. Sanchez, L. Badia, G. Garcia-Belmonte, Y. S. Kang, I. Mora-Sero, J. Bisquert, 2014, J. Phys. Chem. Lett. **5**, 2390-2394.
[5] J. Roa-Rojas, C. Salazar M., D. Llamosa P., A.A. León-Vanegas, D.A. Landínez Téllez, P. Pureur, F.T. Dias, V.N. Vieira, 2008, J. Magn. Magn. Mater. **320**, e104-e106.
[6] H.W. Eng, P.W. Barnes, B.M. Auer, P.M. Woodward, 2003, J. Solid State Chem. **175**, 94-109.
[7] J. Suntivich, K.J. May, H.A. Gasteiger, J.B. Goodenough, Y. Shao-Horn, 2011, Science **334**, 1383-1385.
[8] Y.-H. Huang, R.I. Dass, Z.-L. Xing, Z.-L. Xing, J.B. Goodenough, 2006, Science **312**, 254-257.
[9] H.-L. Liu, H.-C. Hsueh, I.-N. Lin, M.-T. Yang, W.-C. Lee, Y.-C. Chen, C.-T. Chia, H.-F. Cheng, 2011, J. Phys.: Condens. Matter **23**, 225901.

Journal of Physics: Conference Series **614** (2015) 012002 doi:10.1088/1742-6596/614/1/012002

[10] H. Karunadasa, Q. Huang, B.G. Ueland, P. Schiffer, R.J. Cava, 2003, Proc. Natl. Acad. Sci. USA, **100**, 8097-80102.
[11] A.C. Larson, R.B. Von Dreele, General Structure Analysis System, Los Alamos NationalLaboratory Report LAUR, 2000, 86.
[12] Kittel, C., "*Introduction to Solid State Physics*", 8th Ed. University of California, Berkeley 2005.
[13] D.A. Landínez Téllez, D. Llamosa P., C.E. Deluque Toro, Arles V. Gil Rebaza, J. Roa-Rojas, J., 2013, Mol. Struct. **1034**, 233-237.
[14] G. Łupina, J. Dąbrowski, P. Dudek, G. Kozłowski, P. Zaumseil, G. Lippert, O. Fursenko, J. Bauer, C. Baristiran, I. Costina, H.-J. Müssig, L. Oberbeck, U. Schröder, 2009, Appl. Phys. Lett. **94**, 152903.

Production and structural and magnetic characterization of a $Bi_{1-x}Y_xFeO_3$(x = 0, 0.25 and 0.30) system

J A Mejía Gómez[1], GI Supelano García[2], C A Palacio[1], C A Parra Vargas[2],
[1] Grupo GIFAM, Facultad de Ciencias, Universidad Antonio Nariño, Carrera 7 # 21-84, Tunja, Boyacá Colombia
[2] Facultad de Ciencias, Escuela de Física, Grupo de física de materiales, Universidad Pedagógica y Tecnológica de Colombia, Avenida Central de Norte, Tunja, Boyacá, Colombia

E-mail: juliethmejia@uan.edu.co

Abstract. The production and the structural and magnetic characterization of the $Bi_{1-x}Y_xFeO_3$ (x= 0, 0.25 and 0.3) system is reported in this work. The system was produced through the solid-state reaction technique. The morphological characterization obtained by scanning electron microscopy technique evidences the granular behavior. The structural properties were studied by means of X-ray diffraction technique. Magnetization measurements in function of temperature of the $Bi_{1-x}Y_xFeO_3$ (x= 0, 0.25 and 0.3) system were performed with the magnetometer VSM by means of the Zero Field Cooled–Field Cooled method. The results obtained from all the techniques evidence the effect of yttrium on the physical properties of $BiFeO_3$.

1. Introduction

Multiferroic materials exhibit simultaneous ferromagnetic, ferroelectric and/or ferroelastic ordering [1]. Bismuth ferrite ($BiFeO_3$) is one of such materials with antiferromagnetic behavior (T_N = 643 K) and a ferroelectric ordering (T_C = 1103 K), with both properties coexisting at room temperature [2-3]. This compound has ABO_3 Perovskite structure belonging to the R3c space group in which A represents the larger Bi^{3+} ions occupying the cube-octahedral positions and B represents the medium sized Fe^{3+} ions in octahedral coordination [4]. In $BiFeO_3$, the Fe magnetic moments are coupled ferromagnetically within the pseudocubic (111) planes and antiferromagnetically between the near planes; this is called the G-type antiferromagnetic order. Additionally, this compound exhibits a spatial spin-modulated structure (SSMS). The destruction of this SSMS must lead to ferromagnetism and a magnetoelectric effect [2, 5]. In spite of this complex magnetic behavior, this compound has a great potential for information storage, sensors, and electric field controlled devices. It has been shown that by substitution of elements at the A-site (Bi) or B-site (Fe), the SSMS can be suppressed and, therefore, enhancement of the magnetic properties is evidenced [6-15]. Rare-earth elements (La, Nd, Eu, Dy, Gd, Ho, Y, etc.) can be substituted at the A-site and, thus, improve ferroelectric and magnetic properties [7, 8, 12-15]. Guo *et al.* [15] reported that Gd doped $BiFeO_3$ nanoparticles synthesized via sol-gel method increase remnant magnetization with the increase of the Gd substitution. However, the photocatalytic activity was found to decrease as Gd concentration increases. James *et al.* [14] obtained bismuth iron based pigments $Bi_{1-x}M_xFeO_3$ (M= Y, La; x=0, 0.05, 0.1, 0.15, 0.2) by conventional solid state route. X-ray diffraction (XRD) results indicated that a pure phase was formed up to 10 mol% yttrium; for higher concentrations impurity phases are detected. Doping of lanthanum did not strongly affect the lattice parameters due to the similar ionic radii of La^{3+} and Bi^{3+}. In addition, smaller particles were obtained with the substitution of these elements. Magnetic properties of $BiFeO_3$ with substitution of yttrium (10 mol%) were studied by Rao *et al.* [12]. The compounds were obtained via solid-state method. It was suggested that the remnant magnetization was enhanced with the substitution of Y in the A-site due to the suppression of the spin-modulated structure.

Given that the substitution of rare earths is a way of influencing the physical properties of multiferroic BFO, it is relevant to expand research on doping with yttrium concentrations above 10 mol% and establish the effects of this element. In this work, $Bi_{1-x}Y_xFeO_3$ (x= 0, 0.25 and 0.3) ceramics were synthesized by the solid-state route. The results are based on XRD, scanning electron microscopy (SEM) and magnetization studies on the phase formation using two different concentrations of yttrium.

2. Experimental

Bismuth ferrite samples were prepared by means of the conventional solid-state reaction technique, starting from high purity Fe_2O_3, Bi_2O_3, and Y_2O_3 powders. The ground powders were pre-calcined at 700 °C for 2 h in an alumina crucible. The calcined material was ground thoroughly and further calcined at 750 °C for 2 h (the synthesis temperature and time were kept constant for all doping levels) followed by cooling in air by taking the sample out of the furnace. The crystal structure was studied by XRD experiments, with a nickel-filtered Cu-K radiation (λ=1.5406 Å) of an X´Pert PRO PANalitycal diffractometer. Morphological studies of the samples were performed via SEM experiments. For the SEM analysis, the cleaned and dried samples were measured in a JEOL JSM 6490-LVmachine. The magnetization hysteresis (M-H) loop, zero field cooling (ZFC), and field cooling (FC) studies of the $BiFeO_3$ system were carried out by using vibrating sample magnetometers (VSM). Magnetization measurements were performed at several magnetic fields (field range of -3 T – 3 T) and magnetic hysteresis measurements were done at different temperatures, including room temperature (temperature range of 50 - 300 K).

3. Results and discussion

The XRD patterns of the $Bi_{1-x}Y_xFeO_3$ (x= 0, 0.25 and 0.3) samples are shown in Figure 1. It was found that all compounds crystallize in rhombohedral structure with the R3c group, indicating the lack of orthorhombic phases with the yttrium substitution [12]. Besides the main phase BFO, few other peaks in the XRD diffractograms have been observed for secondary phases, for example, that in 32-35° of 2θ. Through the literature review and by means of a Rietveld analysis (performed by GSAS code), the extra peaks observed are $Bi_2Fe_4O_9$.

Changes in the split peaks (012) and (110), typical for $BiFeO_3$, with Y doping suggest the existence of structural effects. The lattice parameters for a $Bi_{1-x}Y_xFeO_3$ system obtained from Rietveld analysis are summarized in Table 1. For the $BiFeO_3$ phase the lattice parameter a is smaller and the lattice parameter c is bigger compared to that reported by other authors [18].

It is important to notice that the results obtained in the present work contradict those obtained in reference [16], which concludes that for higher concentrations of yttrium 10 mol% other Y-phases are formed. In contrast, the content of the impurity $Bi_2Fe_4O_9$ increases as the yttrium concentration increases. In addition, the mechanisms of BFO formation during synthesis process in the presence of yttrium are still unclear.

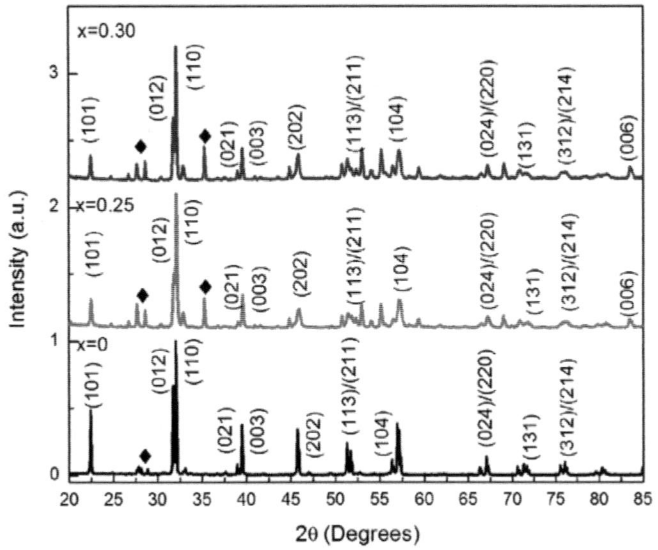

Figure 1. XRD patterns obtained for $Bi_{1-x}Y_xFeO_3$ compounds. The symbol (♦) represents the $Bi_2Fe_4O_9$ phase

The SEM micrographs of the $Bi_{1-x}Y_xFeO_3$ samples are displayed in Figure 2. For x = 0, the grains have a rounded shape and their grain size distribution is estimated to be from 0.5 to about 1 μm (counting around 100 particles). A secondary phase can also be observed (which agrees with the XRD measurements) in a form of agglomerates bigger than 1 μm. With x = 0.25, the rounded shape of the grains has already changed exhibiting a high interconnection of the grains. Such an effect is even stronger with x = 0.30, where bigger agglomerations are observed.

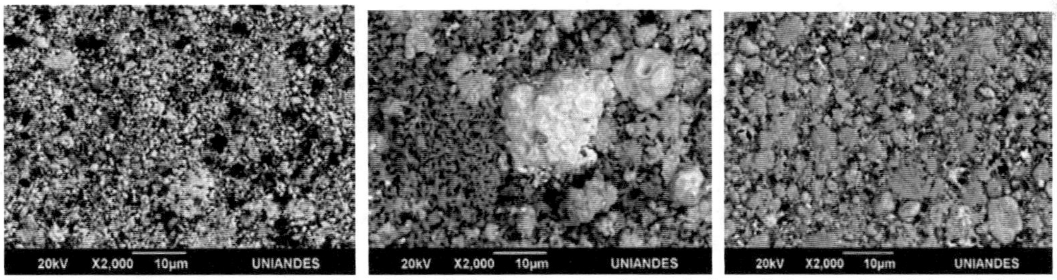

Figure 2. SEM micrographs of the $Bi_{1-x}Y_xFeO_3$ samples. From left to right, x = 0, 0.25, and 0.35

Magnetization hysteresis (M-H) loops taken at room temperature and 50 K are observed in Figure 3. As shown, the conventional loop for BFO at room temperature is very thin, but non-zero remnant magnetization. The values of M_s and M_r are similar for samples reported by Maurya *et al.* [19] synthesized through the same method for the $BiFeO_3$ sample. On the other hand, the samples with the presence of yttrium exhibit higher values for remnant magnetization and lower coercive field, inferring that the substitution of Y in the A-site influences the magnetic properties of BFO systems. The ionic radius of Y^{3+} is smaller compared to that of Bi^{3+}, therefore, doping with this rare earth element affects the bonds between Bi-O and Fe-O. The magnetic behavior of the samples at 50 K (Figure 3 left) is similar to that at 300 K (Figure 3 right) except for saturation magnetization values, which are higher in the first case (Table 1).

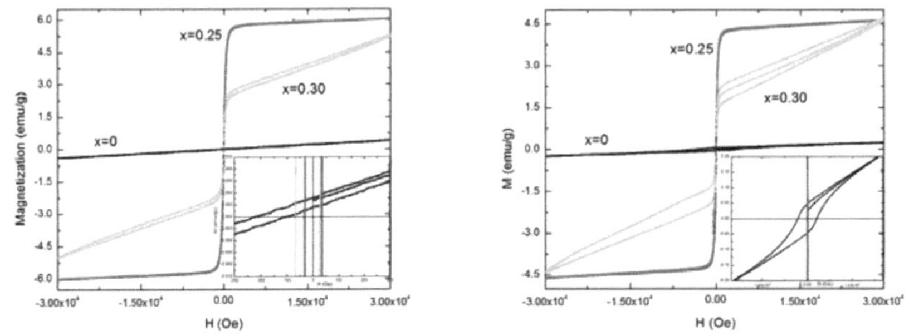

Figure 3. Magnetization hysteresis loops at 50 K (left) and 300 K (right) of $Bi_{1-x}Y_xFeO_3$ samples. Black, red, and green lines correspond to x = 0, 0.25, and 0.30, respectively

Table 1. *a*, *c* lattice parameters and magnetic values obtained for the $Bi_{1-x}Y_xFeO_3$ system

			T = 300 K			T = 50 K		
x	*a* (Å)	*c* (Å)	H_c (Oe)	M_r (Emu/g)	Ms (Emu/g)	H_c (Oe)	M_r (Emu/g)	Ms (Emu/g)
0.00	5.059	13.968	3120	0.05	0.17	212	0.003	-------
0.25	6.519	15.721	11	0.51	6.00	35	0.600	4.60
0.30	6.519	15.722	52	0.90	2.42	54	0.800	2.69

In order to analyze the magnetic ordering in the samples, the temperature dependence of ZFC and FC magnetization curves is presented in Figure 4. The most interesting phenomena are observed in the sample with zero concentration (Figure 4 left). Firstly, the FC and ZFC cycles do not coincide, which implies possible spin glass-like ordering [16]. In addition, comparing these cycles to those obtained by other authors, some differences exist [16-18]. It is possible to notice an anomalous dependence on magnetization temperature, such as a decrease in magnetization that starts at 150 K (which is different from previously reported results), besides being accompanied by a significant difference at about 230 K (this kind of phenomenon was published in ref [18] but it was not explained). One of the reasons for such an anomalous magnetization may be the oxygen vacancy concentration and the existence of Fe^{2+} ions [17], another possible explanation could be the contribution of the impurity phases. In these experiments, it is also noted that the substitution of yttrium affects the magnetic response of BFO as observed in Figure 4 right.

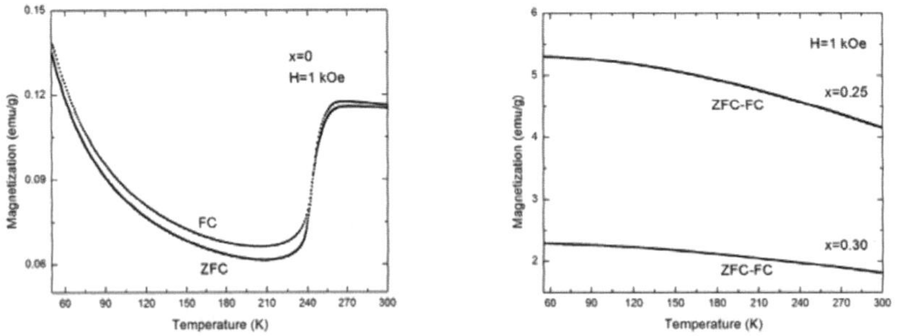

Figure 4. M vs T curves for the samples with zero (left) and with x = 0.25 and 0.30 (right)

4. Conclusions
Bismuth ferrite samples were prepared by conventional solid-state reaction technique. The magnetic responses obtained are typical of ferromagnetic behaviour in a BFO system, affected by the percentage increase of yttrium. The results obtained from XRD and SEM techniques are consistent with changes in the structural properties by Y doping.

5. References
[1] Eerenstein W, Mathur ND and Scott J F 2006 *Nature* **442** 759
[2] Fischer P, Polomska M, Sosnowska I and Szymanski M 1980 *J. Phys. C: Solid State* **13** 1931
[3] Moreau JM, Michel C, Gerson R and James WJ 1971 *J. Phys. Chem. Solids* **32** 1315
[4] Kubel F and Schmid H 1990 *Acta Cryst* **B46** 698
[5] Sosnowska I, Prezenioslo R, Fischer P and Murashov VA 1996 *J Magn Magn Mater* **160** 384
[6] Xu Q, Zhou S, Wen Z, Wu D, Qiu T, Xu M, Potzger K and Schmidt H 2011 *Phys. Let. A* **375** 1209
[7] Kothari D, Reddy V R, Gupta A, Meneghini C and Aquilanti G 2010 *J. Phys.: Condens. Mater* **22** 356001
[8] Maurya D, Thota H, Garg A, Pandey B, Chand P and Verma H C 2009 *J. Phys.: Condens. Mater* **21** 026007
[9] Mathe V L, and Patankar K K 2007 *J. Mater. Sci.* **42** 136
[10] Kim W S, Jun Y K, Kim K H and Hong S H 2009 *J Magn Magn Mater* **321**
[11] Palkar V R, Kundaliya D C, Malik S K and Bhattacharya S 2004 *Phys. Rev B* **69** 212102.
[12] Rao T D, Karthik T and Asthana S 2013 *J. Rare Earth* **31** 370
[13] Puli V S, Pradhan D K, Martínez R, Coondoo I, Panwar N and Katiyar R S 2012 *J. Supercond. Nov. Magn.* **25** 1109
[14] James V, Prabhakar Rao P, Sameera S and Divya S 2014 *Ceram. Int.* **40** 2229
[15] Guo R, Fang L, Dong, W, Zheng F and Shen M 2010 *J. Phys. Chem. C* **114** 21390
[16] Lu J, Günther A, Schrettle F, Mayr F, Krohns S, Lunkenheimer P, and Loidl A 2010 *Eur Phys. J. B* **75** 451
[17] Pradhan A K, Zhang K, Hunter D, Dadson J B, Loiutts G B, Bhattacharya P and Burger A 2005 *J. App. Phys.* **97** 093903
[18] Fruth V, Mitoseriu L, Berger D, Ianculescu A, Matei C, Preda S and Zaharescu M 2007 *Progrg Solid. State Chem..* **35** 193
[19] Maurya D, Thota H, Nalwa K S and Garg A 2009 *J Alloys Compd.* **477** 780

Magnetic Study of Fe_3O_4 Particles for Implementation as Magnetic Fluids

H A Fonseca[1], E Gonzalez[1], C A Parra Vargas[1]
[1]Facultad de Ciencias, Escuela de Física, Universidad Pedagógica y Tecnológica de Colombia, Avenida Central Norte, Tunja, Boyacá, Colombia

E-mail: hugoalexander.fonseca@uptc.edu.co

Abstract. In this work, iron oxide was synthesized by mechanical alloy vibratory disc method. The dependence of magnetic properties on both natural and commercial iron oxide has been studied by using hysteresis loops performed by a vibrating sample magnetometer. The natural oxide phases were determined via X-ray diffraction technique. Magnetic fluid samples were produced by using oleic acid as surfactant and oil as carrier liquid. The viscosity of selected samples was studied at room temperature through rotational viscometer. The results show dependence between the syntheses processes in which the samples were obtained, although experimental results are within those reported in the literature.

1. Introduction

Iron oxide (Fe_3O_4) or magnetite has interesting optical, electric, and magnetic properties for technological applications due to its composition, spinel structure, and ferrimagnetic behaviour. In colloids science, Fe_3O_4 is important in the fabrication of magnetic fluids because the rheological properties can be controlled for application in engineering [1,2] and biomedical [3,4] areas.

These magnetic fluids must have a Newtonian behaviour in the absence of an external magnetic field; high stability against aggregation processes and gravitational sedimentation is achieved by stabilizing the suspension from the particles controlling the densities between the particles and liquid carrier. This can be achieved through steric repulsion (surfactant) [3,4] or by changing the surface chemistry of the particles to obtain electrostatic repulsion between them (peptized) [4,5].

2. Experimental

The natural oxide was obtained from the Hobo – Huila region in Colombia and it was milled in a vibrating disc mill (sample m1). A commercial oxide was acquired from Carbones Andinos (sample m2). These oxides were selected by size and labelled T5, T2 and ST with regards to sift 500, between 200-230 mesh and without sift, respectively. Particle size was determined via optical microscopy

Facultad de Ciencias, Escuela de Física, Universidad Pedagógica y Tecnológica de Colombia, avenida Central norte 39-115, Tunja, Boyaca 15001, Colombia, hugoalexander.fonseca@uptc.edu.co

Journal of Physics: Conference Series **614** (2015) 012004 doi:10.1088/1742-6596/614/1/012004

using Leica digital equipment (DVM200) to measure the smallest diameter of a minimum of 800 particles and adjusted to a lognormal distribution. Qualitative determination of the phases was performed through X-ray diffraction (XRD) using XPERT-PRO PANAlytical equipment and magnetic hysteresis loops by vibrating sample magnetometer (VSM) in a VersaLab at selected temperatures of 300 K and 320 K. Fluid preparation was performed with 20, 30, and 50% of magnetite surfactant oil prepared by mechanical agitation for 2 h at 24 °C and 43 rpm. The viscosities of magnetic fluids were analysed in a rotational viscometer (DV-E1, BROOFIELD) with a spindle of 0.3 without magnetic field.

3. Results and discussions

We found that the T5 and T2 samples were in the 15 – 22 μm and 42 – 47 μm ranges, respectively. For the m1T5 sample, the XDR and size results are shown in Figure 1.

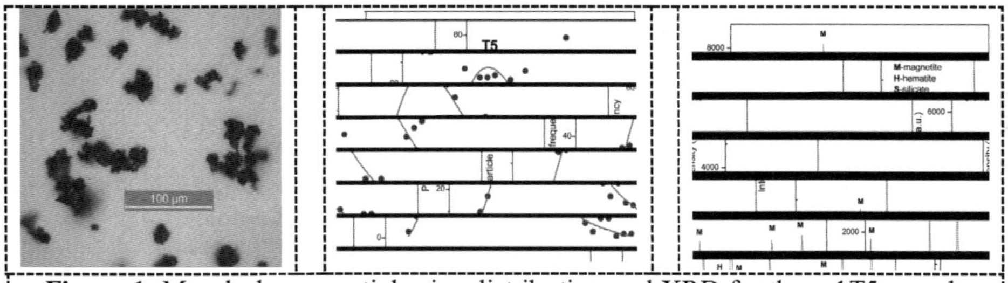

Figure 1. Morphology, particle size distribution and XRD for the m1T5 sample

The irregular particle shape complicates the theoretical study of the magnetic interaction between particles of magnetic fluid, in which a calculation is necessary for the magnetization of the particle in statistical or random manner [6]. With respect to the crystal structure of the m1 sample, the XRD patterns were associated to the stoichiometric phase of magnetite (82%), followed by hematite (17%) and some silica impurities (1%).

The results of magnetic measurements for the ST, T2 and T5 samples showed that saturation magnetization values are different, revealing magnetization dependence on particle size and temperature, respectively (Figure 2).

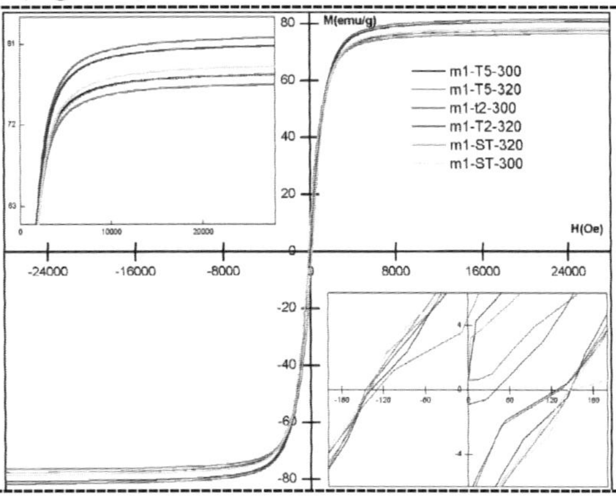

Figure 2. Hysteresis loops for the m1 sample at different temperatures. The upper left and lower right insets are the zoom of the saturation magnetization, and of coercive field, respectively

Journal of Physics: Conference Series **614** (2015) 012004 doi:10.1088/1742-6596/614/1/012004

The dependence on temperature shows the system is in a thermodynamic phase transition associated to dipole moments disorder, which is that thermal energy tends to exceed the energy coupling of the magnetic dipole moments, thus, decreasing the oxide saturation magnetization.

When coercivity increases grain size decreases due to the anisotropy of the particle surface; otherwise, its value would decrease as in the case of nanoparticles [7,8]. The saturation magnetization and coercive field values are shown in Table 1 for the m1 and m2 samples because the micrometric size of the particles present multidomain parameters.

Table 1. Saturation magnetizations (Ms) and coercive field (Hc) values

Sample	T (K)	Ms (emu/mol)		Hc (Oe)	
		m1	*m2*	*m1*	*m2*
ST	300	78.42	84.42	142.5	51.55
	320	77.55	81.5	142.5	51.02
T2	300	81.59	83.95	132.2	53.07
	320	80.7	83.07	133.4	48.5
T5	300	77.53	84.34	144.62	71.05
	320	76.45	83.72	147	69.27

Table 1 shows that saturation magnetization for m1 is lower than for the m2 sample, respectively, due to the presence of the hematite and silica phase. This decrease in saturation magnetization is not significant to affect the magnetic response of the sample since their values are within the ranges reported in the literature [6-8]. The difference between the m1 and m2 samples is notable in terms of the coercive field obtained, attributed to the percentage ofhematite m1 present in the sample. In this case, the m2 sample can be a better candidate to study rheological or structural properties (Figure 4) of magnetic fluid under the action of a small external field.

Viscosity measurements for the m1 and m2 samples are given in Figure 3. The different curves correspond to the fast sedimentation in sample m2 and stability in sample m1.

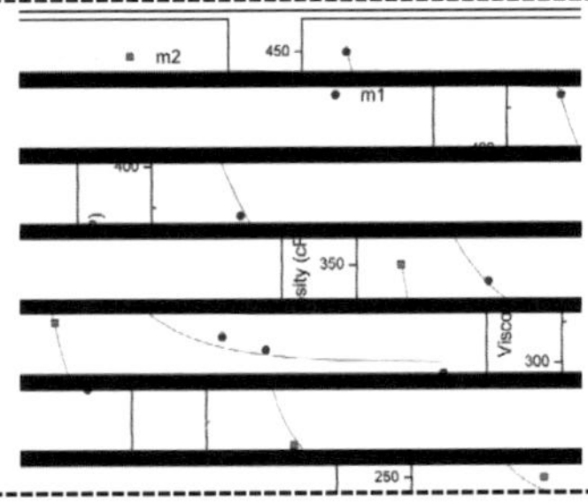

Figure 3. Viscosity of magnetic fluids elaborated from the m1 and m2 samples, as a function of revolutions per minute

The rheological behaviour of fluids changes in the presence of an applied external magnetic field, reaching threshold stress and high viscosity. The particles align their magnetic dipole moments with the magnetic field direction and form different structures in this direction, which increases flow resistance [9]. Figure 4 shows the structures produced for different magnetic field orientations. We see in the images that a portion of iron oxide particles were condensed, forming chains or columns, depending on the direction of the applied magnetic fields. These phenomena can bring up the feasibility of electro-optical devices by using homogeneous ferrofluid films [10].

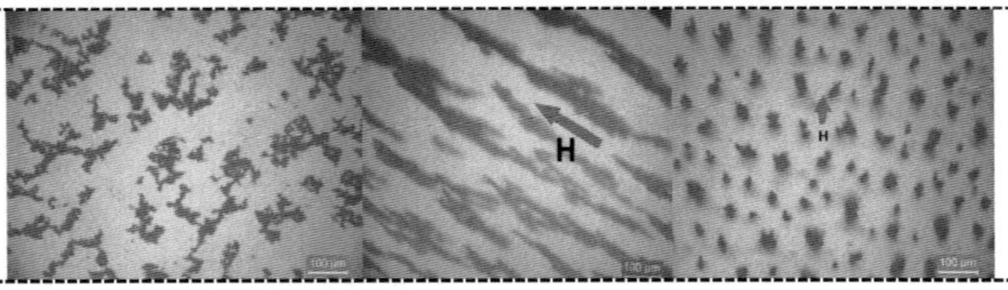

Figure 4. Formations of magnetic structures for the m1 sample. a) without magnetic field, with parallel magnetic field (b), and normal (c) to the surface

When magnetic field ceases, the magnetic fluid tends to its original disordered state with a quasi-reversible process appearing, having a tendency for paramagnetic behaviour, which is due to the presence of the surfactant coating that covers the particle with a film, moderating the molecular interaction through steric repulsion.

4. Conclusions

The magnetic properties of Fe_3O_4 show dependence on the synthesis process, particle size, and temperature. These oxides were dispersed in carrier liquid with magnetorheological properties. The structures formed in the medium depend on the direction of the magnetic field and obtained fibrous or cylindrical structures. This behaviour with the magnetic field enables their rheological control for applications in engineering.

Acknowledgments

The authors thank the *Centro de Gestión de Investigación y Extensión* (CEDEC) of the faculty of Engineering, call 001 for research groups.

5. References

[1] Donado F., Mendoza, C.Y., Valencia, R., Santiago Garcia, J.A., 2007. Estudio experimental de la viscosidad de un fluido magneto-reológico a base de magnetite mineral, *Rev. Mex. Fís.* E **53** 31 – 40.
[2] Friák M., Schindlmayr A., Scheffler M., 2007. Ab initio study of the half-metal to metal transition in strained magnetite, *New. J. Phys.* **9** 1 – 15
[3] Shokrollahi H., 2013. Contrast agents for MRI, *Mater. Sci. Eng.* C **33** 4485 – 4497
[4] Qu J., Liu G., Wang Y., Hong R., 2010. Preparation of Fe3O4-chitosan nanoparticles used for hyperthermia, *Adv. Powder Technol.* **21** 461 – 467.
[5] Betancourt R., Saldívar R., Rodríguez Fernández O., 2004. Preparación de látices magnéticos, *Rev. Mex. Fís.* **50** 33 – 36.
[6] Shaw D. J., 1970. *Introducción a la química de superficies y coloides* (Madrid:Alhambra) **Ch1**
[7] Blanco-Mantecón M., O'Grady K., 2006. Interaction and size effects in magnetic nanoparticles, *J. Mag. Magn. Mater.* **296** 124 – 133.
[8] Özdemir Ö., Dunlop D. J., Moskowitz B. M., 2002. Changes in remanence, coercivity and domain state at low temperature in magnetite, *Earth planet. sci. Lett.* **194** 343 – 358.
[9] Bronstein L. M., Huang X., Retrum J., Schmucker A., Pink M., Stein B. D., Dragnea B., 2007. Influence of iron oleate complex structure on iron oxide nanoparticle formation, *Chem. Mater.* **19** 3624 – 3632.
[10] Yang S. Y., Horng H. E., Hong C. Y., Yang H. C., 2002. Structures, Optical Properties and Potentially Electro-Optical Applications of Magnetic Fluid Films, *J. Sci. Eng.* **5** 85-93

Quantum magnetism in spin-3/2 chains

J. J. Hernández-Sarria[1], A. Argüelles[1], K. Rodríguez[1]

E-mail: karem.c.rodriguez@correounivalle.edu.co
[1]Departamento de Fśica, Universidad del Valle, AA25360 Cali, Colombia

Abstract. Herein, we characterize the quantum phases present in spin-3/2 fermionic chains. This system may be achieved by ultracold gases loaded onto one-dimentional optical lattices in the hard-core regime. The physics of the lowest band in general is described by the Hubbard Hamiltonian that becomes a generalized Heisenberg model in this strongly-correlated regime. By performing calculations of the energy spectrum, correlation functions, level spectroscopy and chirality we analyze the Mott-insulator phases in absence and presence of an external quadratic Zeeman field.

1. Introduction

Ultracold gases in optical lattices constitute an extraordinary toolbox for the analysis of strongly-correlated gases under extremely well-controlled conditions [1, 2], as highlighted by the observation of the superfluid to Mott-insulator (MI) transition in ultracold bosons [3], recently followed by the experimental realization of the metal to MI transition in two-component fermions [4, 5]. Optical traps permit simultaneous trapping of various Zeeman sublevels, allowing for multi-component (spinor) formed by atoms with non-zero total spin F. They constitute an ideal system to investigate the interplay between external and internal (spin) degrees of freedom. As a result of this interplay, the ground state physics is very rich, providing novel scenarios for quantum magnetism. Introducing stronger interactions and smaller occupation number open the fascinating possibility of several insulating phases according to different spin correlations. Manipulation of spins and magnetism has been a motivating goal in a great deal of condensed matter, atom optics and quantum information research. Hence, spinor atoms in optical lattices are good candidates for the achievement of strongly-correlated quantum magnetic systems [6].

Spinor fermions have become the focus of rapidly growing interest, initially motivated by experiments on condensation-superfluidity crossover in two-component fermions [2]. Moreover, the proximity to the Mott insulating phase in fermionic systems is the origin of many intriguing phenomena in condensed matter physics [7]. Such systems are perfect quantum simulators of the Hubbard model and, thus, shed light upon the problem of high T_c superconductivity. This has been proposed for $F = 1/2$ [8] and recently considered with three-component fermions [9]. Interestingly, the suppression of conductivity in the system is a result of the interactions and not the consequence of a filled Bloch band like in the usual insulators.

High spin on cold-atomic systems with high symmetries present strong quantum fluctuations [10], rare in condensed matter and worth exploring. Currently, relatively few works have been conducted for high spin fermions, despite their interesting properties.

In this paper, we characterize repulsively interacting spin-3/2 fermions in the Mott-insulator regime. Contrary to spin-1/2, the MI of spin-3/2 presents two distinct magnetic phases in one dimension (1D) and absence of an external field: a gapless spin liquid (SL) and a gapped dimerized spin Peierls (SP) phase, depending on the inter-atomic interactions. Spin-3/2 fermions are the smallest spin fermionic systems that exhibit spin-changing collisions and, hence, sensitivity to the quadratic Zeeman effect (QZE), or single-ion anisotropy. The QZE role must be taken into account to study high spin systems [11]. Therefore, a detailed analysis with and without an external magnetic field is performed by studying the spectrum, correlations, and level-crossing spectroscopy providing strong numerical proof of the quantum nature of the Mott-phases involved.

2. Effective spin-3/2 Hamiltonian with super-exchange interactions

We consider a balanced mixture of spin-3/2 fermions in a 1D lattice with zero total magnetization ($\mathcal{M} = 0$). For a deep lattice and low filling (single-band regime) the most generic Hamiltonian describing 4-component fermions with equal masses interacting via contact potential is given by the Hubbard model. The interaction strengths $g_{0,2} = 4\pi\hbar^2 a_{0,2}/M$ characterize the s-wave channels with total spin $\{0, 2\}$ (the only available due to symmetry), with $a_{0,2}$ the scattering lengths and M the atomic mass. Although, typically a_0 and a_2 are similar, their values may be controlled by means of microwave dressing [12] or optical Feshbach resonances [13]. Along this paper, we use the following change of variables (following Ref [11]), $G = (g_0 + g_2)/2$ and $g = (g_2 - g_0)/(g_2 + g_0)$, where g gives a measurement of the spin-changing collision strength. The interactions preserve $\mathcal{M}$ and the linear Zeeman effect induced by a magnetic field does not play any role. However, due to spin-changing collisions which re-distribute the populations of the different components while preserving $\mathcal{M}$, the QZE becomes crucial in spinor gases. The field is parametrized by the externally tunable constant q and is experimentally controllable by means of a magnetic field or microwave or optical dressing.

For large-enough interactions and for a chemical potential μ larger than a critical value of $\mu_c(t)$, the system enters into the MI phase with maximally one fermion per site. Under these conditions, the charge degrees of freedom are frozen and the spin degrees of freedom are described by an effective Hamiltonian with super-exchange interactions, which can be calculated via perturbation theory in the hopping term (t). The resulting Hamiltonian is of the form $\hat{H} = \hat{H}_0 + \hat{H}_{sc}$, where $\hat{H}_0$ contains self-energies and spin preserving interactions, whereas $\hat{H}_{sc}$ has the spin-changing interactions. In the following we will use a renormalized QZE parameter $\tilde{q}$ defined as $\tilde{q} = qG/t^2$.

3. Phases in the Mott regime without external fields

When the charge gap opens, the insulating ground state phases are located in the repulsive interaction regime where $g_{F=0,2} > 0$. In terms of g and G, when $q = 0$ [14, 15], the ground-state of the system for $-1 < g \leq 0$ is a gapless spin liquid phase with three gapless spin-modes, including the exactly solvable $SU(4)$ point ($g = 0$) [16]. For $0 < g < 1$ the ground state is a spin-Peierls phase, which exhibits a spin gap and long-range dimerization order. At $g = 0$, the system undergoes a Kosterlitz-Thouless-like transition between the two phases [17].

Energy spectrum. The system energy is plotted as a function of the energy levels in Fig. 1 for several values of the spin-changing strength $g = \{-0.5, 0, 0.5\}$ and for system size $L = 6$. All the main features of the magnetic phases are present in this small system and the results are retrieved by using both full diagonalization and the Lanczos algorithm. One can observe the excited states energies rising with the index level (N), following a linear tendency with a non-universal slope in both phases and boundary conditions (BC). For both periodic BC (PBC) panels (a-c) and open BC (OBC) panels (d-f) the decay-rate energy of the spin liquid phase (panels (a-b) and (d-e)) is half smaller than the respective in the dimerized phase (panels (c) and (f)), showing that in the latter this decay is faster including an energy jump given by its characteristic gap.

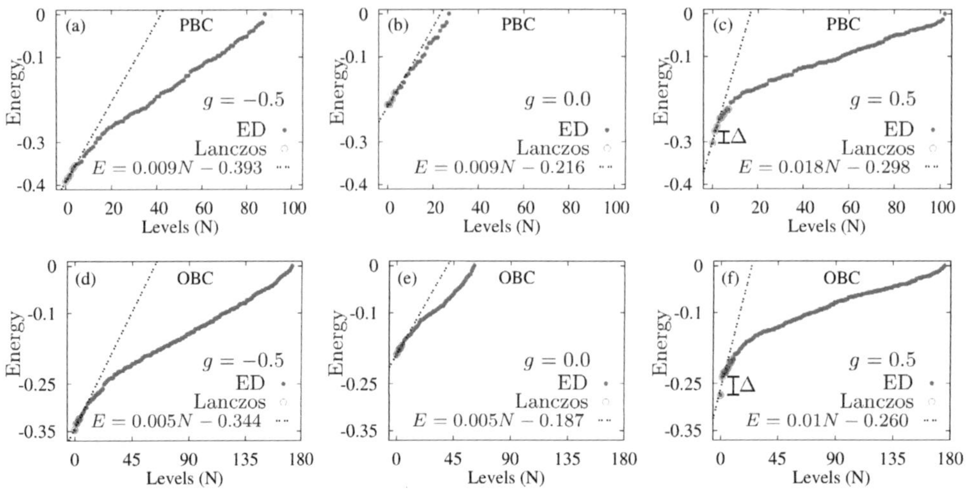

Figure 1. Energy spectrum for $L = 6$ lattice sites without magnetic field for several values of g using PBC (a) and (b) in the SL phase and (c) in the SP phase. And using OBC (d) and (e) in the SL phase and (f) in the SP phase.

Energy gap. The energy gap decays in system size for the dimerized phase. Its behavior is also not universal and depends on the BC. When PBC are imposed, the ground-state is two-fold degenerated in the thermodynamic limit. For OBC, the energy difference is exponentially small in system size. Figure 2 shows that the gap decays exponentially for both periodic (a) and open (b) BC and the values are $\Delta_{\infty,PBC} = 0.016$ and $\Delta_{\infty,OBC} = 0.011$, respectively. Remarkably, the calculations were performed by using the Lanczos algorithm with no more than 10 lattice sites, showing the robustness of this gap.

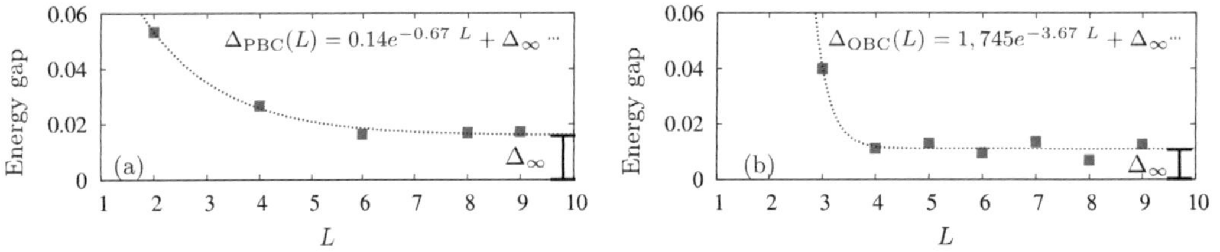

Figure 2. Gap-decay in the SP phase using both periodic (a) and open (b) BCs with thermodynamic-limit gap-values $\Delta_{\infty,PBC} = 0.016$ and $\Delta_{\infty,OBC} = 0.011$, using $g = 0.5$.

4. Field-induced Mott phases

Isotropic Heisenberg antiferromagnet (iHAFM) at $|\tilde{q}| \gg \tilde{q}_c$. In the limiting case, for a very strong magnetic field, the degeneracy among $\pm 1/2$ and $\pm 3/2$ is lifted, the spin changing super-exchange H_{sc} is negligible, and the two spin manifolds separate. Second-order perturbation theory projects the 4- to a 2-components iHAFM Hamiltonian, favoring one manifold depending on the sign of q ($q < 0$ for $\pm 3/2$ else $\pm 1/2$), see definitions in [11]. Interestingly, the QZE induces phase transitions between the two $q = 0$ phases and the iHAFM.

Dimerized phase vs. iHAFM phase ($g > 0$). In the SP phase, the translational symmetry by one site is spontaneously broken, being a characteristic of 1D systems. In the ground state, neighboring pair of spins with antiferromagnetic interaction tends to form a spin singlet. This arrangement is studied by the dimer-order parameter $\hat{D}_i = (-1)^i \vec{S}_i(\vec{S}_{i-1} + \vec{S}_{i+1})$ and the long-range order (LRO) in the dimer-dimer correlation (DDC) function $f_0 = \lim_{\delta\to\infty}\langle \hat{D}_i \hat{D}_{i+\delta}\rangle$.

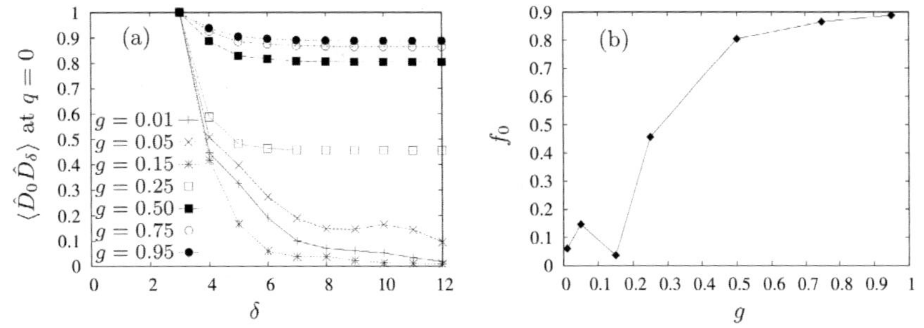

Figure 3. (a) DDC spatial function and (b) LRO f_0 as a function of g at $q = 0$.

The calculations were done by using the Density-Matrix Renormalization Group (DMRG) method [18], with up to $L = 36$, open BC, and matrix dimension 20. Figure 3 shows a stronger LRO as the interaction $g > 0$ increases in absence of QZE. In 3(a), the DDC is presented as a function of δ, while the panel (b) shows the rise of f_0 with the magnitude of the spin-changing exchange g.

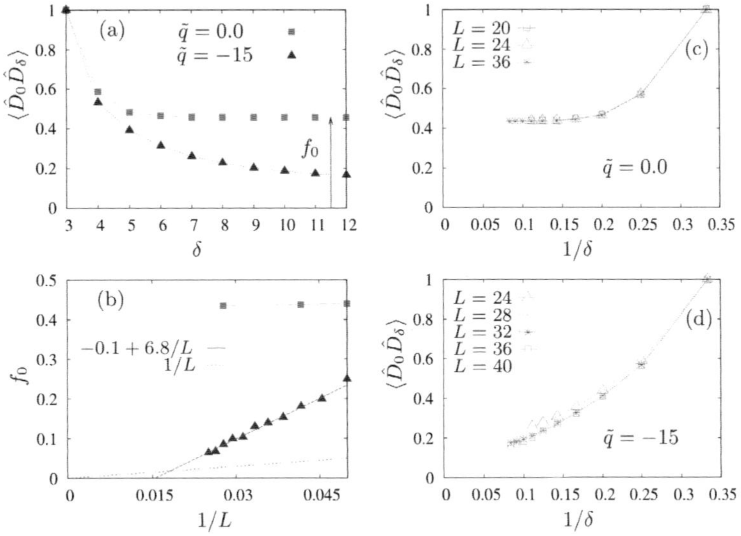

Figure 4. (a) DDC for $\tilde{q} = 0$ and $\tilde{q} = -15$. (b) Finite-size scaling $f_0 \to 0$ for $\tilde{q} = -15$ and f_0 is finite for $\tilde{q} = 0$. (c) DDC for several system sizes for $\tilde{q} = 0$ and (d) for $\tilde{q} = -15$.

In figure 4(a), we observe that the long-range order disappears in the Heisenberg phase, given that the spatial DDC for $\tilde{q} = -15$ decays faster compared with the $q = 0$ case. A finite size scaling has been performed (4(b)), showing the fast decay of f_0 to lower values of $1/L$ for the iHAFM (filled triangles), while f_0 goes almost constant for the SP (filled squares). Panels (c) and (d) show the DDC as a function of $1/\delta$. In (c), the correlation function tends to a finite value for $q = 0$, whereas it tends to zero when $\tilde{q} = -15$ in panel (d). The decay of the spin-spin correlation (SSC) function $\langle \vec{S}_0 \vec{S}_\delta \rangle$ is shown in Fig. 5(a). The corresponding exponential fit is presented in 5(b) for the SP phase and the algebraic fit is shown in panel (c) for iHAFM, elucidating the 1D counterpart of the 3D Néel order. Panels (d) and (e) present the SSC as a function of $1/\delta$ for several system sizes, showing fast decay in the dimerized phase at $q = 0$ and the $1/\delta$ decay expected for the iHAFM phase.

Level spectroscopy. In a finite chain, the two ground states of the dimerized phase split into a unique ground and an excited state separated by an energy gap exponentially small in the

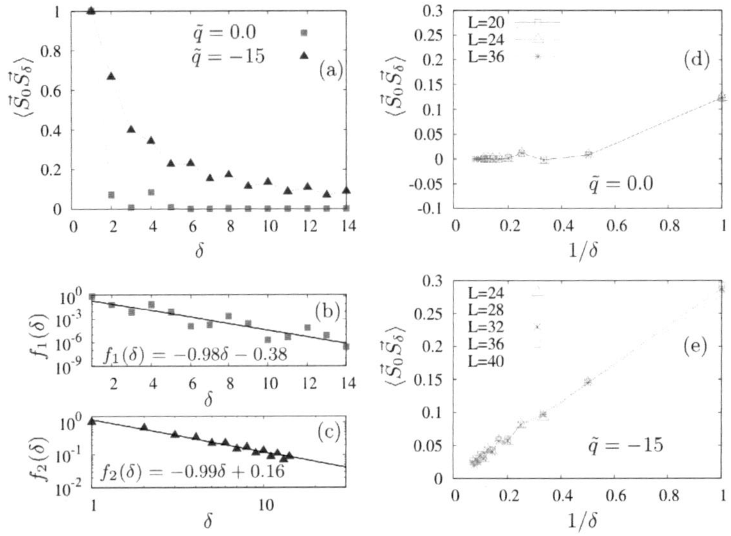

Figure 5. (a) SSC for $\tilde{q} = \{0, -15\}$. (b) Shows the exponential decay of SSC for $\tilde{q} = 0$ and (c) the algebraic decay for $\tilde{q} \gg 1$. (d) SSC for several system sizes for $\tilde{q} = 0$ and (e) for $\tilde{q} = -15$.

system size for $0 > |\tilde{q}| > |\tilde{q}_c|$. In contrast, the lowest excited state in the iHAFM ($|\tilde{q}| > |\tilde{q}_c|$) is a degenerate triplet. Lanczos results show a clear excited singlet-triplet level crossing and the crossing points up to 14 sites with PBC are shown in Fig. 6(a). A finite-size scaling (panel (b)) determines the phase transition line in the thermodynamic limit, L_∞, curve in (a), specifying the $\tilde{q}_c$ values, suggesting a Kosterlitz-Thouless phase transition between the two phases. For large g, although with non-negligible corrections, $\tilde{q}_c$ follows a $1/L^2$ extrapolation like in the $J_1 - J_2$ AFM chain [19]. For $g \ll 1$ finite-size effects prevent any reliable extrapolation law.

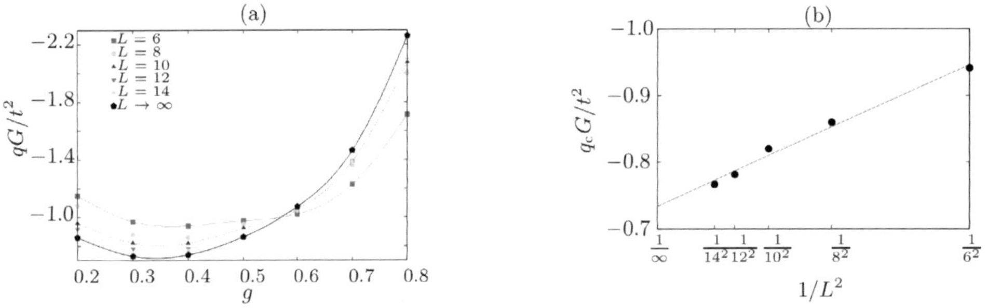

Figure 6. (a) Level-crossing spectroscopy for fixing the phase transition boundary for several system sizes. (b) Finite-size scaling example.

Spin liquid phase vs. iHAFM ($g \leq 0$). In this case, a phase transition is found between these two gapless phases. Due to the experimental relevance, we monitor in figure 7 (a) the chirality as a function of the QZE, $\tau = \frac{1}{L}[(N_{3/2} + N_{-3/2}) - (N_{1/2} + N_{-1/2})]$, ranging from 0 ($q = 0$) to 1 in iHAFM, using the DMRG method. τ is conserved only in the absence of the field. On the other hand, the particle number is conserved in each manifold, but the difference between them is not; a fact reflected in the quasi-saturation behavior. The phase transition cannot be identified with τ alone, so we study the critical exponent γ of the SSC. Given that, for $g < 0$ the correlation $\langle \hat{S}^z_i \hat{S}^z_j \rangle$ oscillates with 4-site periodicity, the bulk correlation envelope function must be studied [20] instead. Thus, we first calculated the bulk SSC function given by $\langle \hat{S}^z_i \hat{S}^z_{i+j} \rangle_{\text{bulk}} \equiv \frac{1}{4}\sum_{k=0}^{3} \langle \hat{S}^z_{i+k} \hat{S}^z_{i+k+j} \rangle$ and later obtained the γ from its envelope decay. This gives an abrupt jump of γ to -1 at the phase transition (figure 7(b)), preserving the main feature of $g = 0$ for small enough $|g|$. In the iHAFM, the magnetic order is more favorable due to the enhanced antiferromagnetic spin correlations $\langle \hat{S}_z^i \cdot \hat{S}_z^j \rangle \propto |j - i|^\gamma$ with $\gamma = -1$. A jump in γ at $\tilde{q}_c$ confirms the commensurate-incommensurate nature of the phase transition. γ is relevant only for phases with algebraically decaying correlations, which excludes the SP phase.

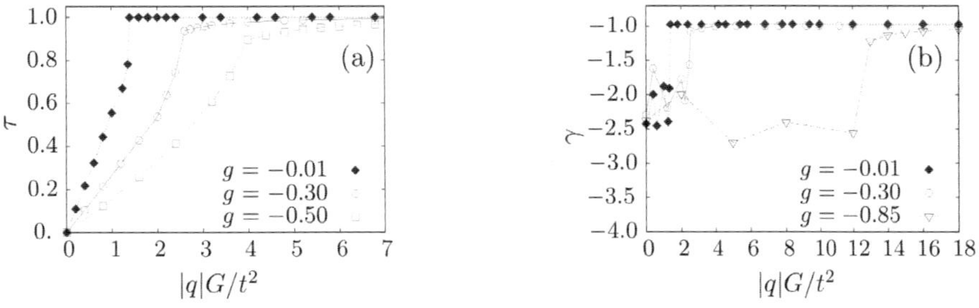

Figure 7. (a) Chirality τ and (b) critical exponent γ as a function of $\tilde{q}$ for $g < 0$.

5. Conclusions

We studied the Mott-insulator phases of spin-3/2 fermions in the absence and presence of an external QZE. We described the pseudo-spin-1/2 iHAFM phase obtained for large enough fields. By decreasing the QZE, this phase undergoes, depending on the scattering lengths, through either a Kosterlitz-Thouless transition into a gapped dimerized phase or a commensurate-incommensurate transition into a gapless spin liquid. In particular, 4-component fermions constitute a unique scenario for experiments on a field-induced commensurate-incommensurate transition. They are the smallest spin fermionic systems that exhibit spin-changing collisions and, hence, sensitivity against QZE.

Acknowledgments

AA and KR thank Luis Santos and Temo Vekua for their support and fruitful discussions. This work has been partly supported by Universidad del Valle under the project CI-7919. KR thanks the DFG for support (GRK665) and the Center of Excellence for Novel Materials (CENM) at Universidad del Valle for financial support for the research group.

References

[1] Lewenstein M, Sanpera A, Ahufinger V, Damski B, Sen A and Sen U 2007 *Advances in Physics* **56** 243–379
[2] Bloch I, Dalibard J and Zwerger W 2008 *Rev. Mod. Phys.* **80** 885–964
[3] Greiner M, Mandel O, Esslinger T, Hansch T W and Bloch I 2002 *Nature (London)* **415** 39
[4] Jördens R, Strohmaier N, Günter K, Moritz H and Esslinger T 2008 *Nature (London)* **455** 204
[5] Schneider U, Hackermuller L, Will S, Best T, Bloch I, Costi T A, Helmes R W, Rasch D and Rosch A 2008 *Science* **322** 1520–1525
[6] Jrgensen O, Heinze J and Lhmann D S 2013 *New Journal of Physics* **15** 113017
[7] Jaramillo J, Greschner S and Vekua T 2013 *Phys. Rev. A* **88**(4) 043616
[8] Hofstetter W, Cirac J I, Zoller P, Demler E and Lukin M D 2002 *Phys. Rev. Lett.* **89** 220407
[9] Paananen T, Martikainen J P and Törmä P 2006 *Phys. Rev. A* **73** 053606
[10] Wang D, Li Y, Cai Z, Zhou Z, Wang Y and Wu C 2014 *Phys. Rev. Lett.* **112**(15) 156403
[11] Rodríguez K, Argüelles A, Colomé-Tatché M, Vekua T and Santos L 2010 *Phys. Rev. Lett.* **105**(5) 050402
[12] Papoular D J, Shlyapnikov G V and Dalibard J 2010 *Phys. Rev. A* **81** 041603
[13] Fedichev P O, Kagan Y, Shlyapnikov G V and Walraven J T M 1996 *Phys. Rev. Lett.* **77** 2913–2916
[14] Lecheminant P, Boulat E and Azaria P 2005 *Phys. Rev. Lett.* **95** 240402
[15] Wu C 2005 *Phys. Rev. Lett.* **95** 266404
[16] Sutherland B 1975 *Phys. Rev. B* **12** 3795–3805
[17] Wu C 2006 *Mod. Phys. Lett. B* **20** 1707–1738
[18] Verstraete F and Cirac J I 2006 *Phys. Rev. B* **73** 094423
[19] Okamoto K and Nomura K 1992 *Physics Letters A* **169** 433 – 437
[20] Yamashita Y, Shibata N and Ueda K 1998 *Phys. Rev. B* **58** 9114–9118

Spin polarization in non-magnetic nanostructures

JA Zúñiga[1], ST Pérez Merchancano[2], and LE Bolívar Marínez[2]

[1] Department of Mathematics, Universidad del Cauca, Popayán, Calle 5 # 4-70, Colombia
[2] Department of Physics, Universidad del Cauca, Popayán, Calle 5 # 4-70, Colombia

E-mails: jazuniga@unicauca.edu.co, sperez@unicauca.edu.co

Abstract. Quantum transport that takes into account spin polarization has a high potential for research on optimal heterostructures for fabrication of nano-spintronic devices and quantum computing. This work theoretically analyzed different materials on the basis of type-II strained heterostructures like InAs/GaSb, InSb/GaSb, InSb/GaAs, and GaSb/GaAs by means of the spin polarization that considers the interacting spin coupling type: k^3-Dresselhaus and Rashba in the electric barriers and the influence of in-plane magnetic field on spin polarization. The results obtained for the polarization of spin are in function of the energy applied to the electron, well width, height of barrier, and magnetic field intensity. This suggests that these features could be engineered in the fabricating of tunable spin-dependent electronic devices, such as spin switches based on double-barrier non-magnetic semiconductors.

1. Introduction

Spin-dependent electron transport in low-dimensional mesoscopic systems has recently drawn much attention due to its potential for future electronic device applications in the field of spintronics [1-3]. This new emerging field deals with the active manipulation of the electron's spin (and not only its charge), adding a spin degree of freedom to the conventional charge-based technology, which has the potential advantages of multifunctionality: longer decoherence times and lengths, increased data processing speed, decreased electric power consumption, and increased integration densities compared with conventional semiconductor devices [1]. In addition to improving contemporary technology, spintronics may also contribute to the field of quantum computation and quantum information in which the quantum information may be contained in the spin, transmitted when the spin is polarized [4].

In the field of spintronics, one of the most important requirements is the completion of spin injection. Hence, some studies have focused on this topic [5-8], particularly, in ferromagnetic (FM) metals in semiconductor heterostructures, but the efficiency of spin injection is disappointingly small due to the conductivity mismatch between the FM metal and the semiconductor. Therefore, an alternative approach is based on the combination of an electric barrier (EB) and a magnetic barrier (MB) in a two-dimensional electron gas system (2DEG) proposed by Papp and Peeters [9,10]. The effects caused by the MB system are very different from the well-known potential barrier because the electron tunneling becomes a two-dimensional (2D) problem. Transmission depends not only on the energy of the impeding electrons, but also on the direction of their velocity toward the barrier. When it takes into account electron charge and spin, electronic transport can become spin polarized. Accordingly, in this work we only considered heterostructures based on double-barrier non-magnetic semiconductors in the presence of EB, as proposed by JD Lu *et al.* [11] and magnetic field in the plane of the barrier, also of the interacting spin coupling type: k^3-Dresselhaus and Rashba in the barriers that according to Glazov *et al.* [12], who investigated the effect of spin-dependent resonant tunneling through the symmetric double barrier in the presence of the Dresselhaus spin-orbit (DSO) term and found that the transmission coefficient was strongly dependent on the spin orientations and on the wave vector of the electrons. Also, it has been found that the Rashba spin-orbit interaction (RSO) plays a more important role than the DSO term in the large spin polarization at zero magnetic field [11]. Therefore, it is expected that said behaviors are similar when spin-dependent tunneling is analyzed through the double-barrier heterostructure with the influence of in-plane magnetic field and the effects of Dresselhaus and Rashba spin-orbit coupling.

2. Model and theoretical calculation

Analysis of electron tunneling initially considered a wave vector $\vec{k}=\left(k_{\parallel}, k_z\right)$ through a double-barrier symmetrical potential height V_0 of type zinc-blende based on double-barrier non-magnetic semiconductors grown along the $[001]$plane (Figure 1). The components wave vector $\vec{k}$, characterized in that, the first is in the plane of the barrier and the following is a normal component to the direction of tunneling. Indeed, $\vec{k}=\left(k_{\parallel} Cos\varphi,\, k_{\parallel} Sin\varphi,\, k_z\right)$, where φ is the polar angle of the wave vector $\vec{k}$ in-plane $[100]$ and $[010]$.

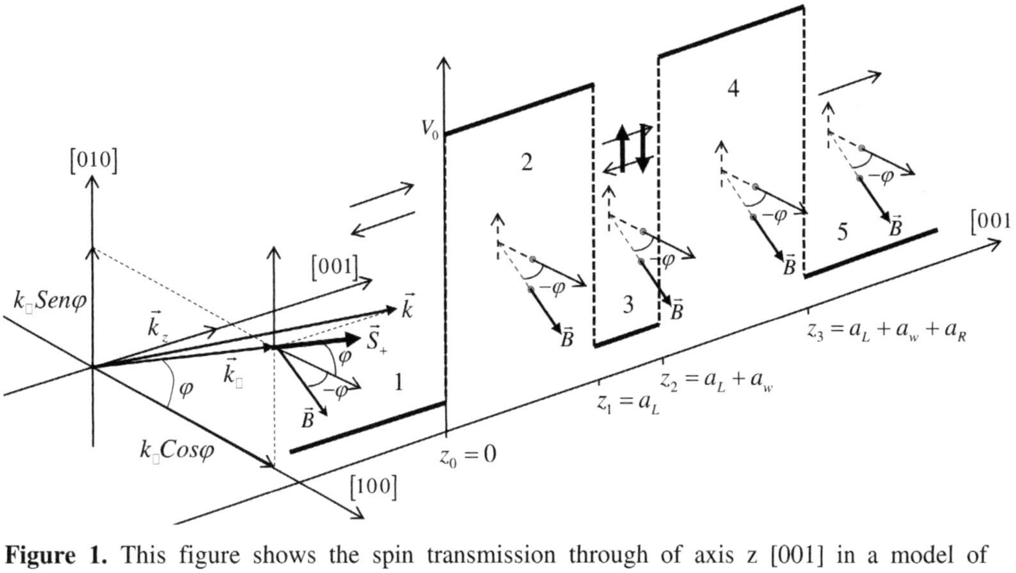

Figure 1. This figure shows the spin transmission through of axis z [001] in a model of symmetrycal electric barrier-double, under the influence of in-plane magnetic field $\vec{B}$ on spin polarization. $a_{L,R}$ are the barrier width and a_w is the well width.

By calculating spin polarization and the probability of tunneling for up and down spin under the RSO and DSO, considering influence of in-plane magnetic field, we can build the Hamiltonian:

$$\hat{H}_j=-\frac{\hbar^2}{2\,m_j^*}\left(k_z^2+k_{\parallel}^2\right)+V_0+\hat{H}_D+\hat{H}_R+\hat{H}_Z \tag{1}$$

where m_j^* the effective mass, with $j=w,b$ (well and barrier), V_0 is the potential of EB, $\hat{H}_D$ is the spin-dependent k^3-Dresselhaus term is simplified [14] to $\hat{H}_D=\beta_D\left(\hat{\sigma}_x k_x-\hat{\sigma}_y k_y\right)k_z^2$, $\hat{H}_R$ is in-plane Rashba term due to heteropotential asymmetry $\hat{H}_R=\alpha_R\left(\hat{\sigma}_x k_y-\hat{\sigma}_y k_x\right)$, where β_D and α_R are constants characteristic of the material and $\hat{H}_z$ Zeeman spin splitting is related to in-plane magnetic field, given by the expression $\hat{H}_Z = \frac{1}{2} g^*_j \mu^j_B (\hat{\sigma} \cdot \vec{B})$. Where $\hat{\sigma} = (\sigma_x, \sigma_y, \sigma_z)$ and each $\hat{\sigma}_i$ as Pauli matrices, k_i the components of the electron wave vector, g^*_j corresponds to effective Landé factor, $\mu^j_B = e\hbar / 2m^*_j$ is the Bohr magnetron and $\vec{B} = (BCos\theta, BSin\theta, 0)$ magnetic field orientation.

Solving the Schrödinger-Pauli equation, $\hat{H}_j \hat{\psi} = E\hat{\psi}$, generated by the expression (1) requires that the wave function, $\hat{\psi}$, is disengaged for the spin and magnetic field, considers the spinors that diagonalize the terms: Dresselhaus $\hat{\chi}^D_{\pm}(\varphi) = \frac{1}{\sqrt{2}}(1, \mp e^{-i\varphi})^T$, and Zeeman $\hat{\chi}^B_{\pm}(\theta) = \frac{1}{\sqrt{2}}(1, \pm e^{-i\theta})^T$, and a function $u = u_{\pm}(z)$ that describes the behavior of the electron in the tunneling direction and a plane wave and parallel to the plane of the EB $\exp(i\vec{k}_{\parallel} \cdot \vec{\rho})$ in which $\vec{\rho}$ is a vector in the plane of the barrier given by $\vec{\rho} = (x, y)$, consequently,

$$\hat{\psi} = \hat{\chi}^B_{\pm} u_{\pm}(z) \exp(i\vec{k}_{\parallel} \cdot \vec{\rho}) \tag{2}$$

Under the considerations of the wave function, $\hat{\psi}$, the symbol $\sigma = \pm$ refers to the states of spin up $|+\rangle$ and spin down $|-\rangle$. Accordingly, the function u_σ is defined in the regions 1, 2, 3, 4, and 5 as:

$$u_\sigma(z) = \begin{cases} A^1_\sigma e^{ik_\sigma(z-z_0)} + B^1_\sigma e^{-ik_\sigma(z-z_0)} & if \quad z < z_0 \\ A^2_\sigma e^{i\rho_\sigma(z-z_0)} + B^2_\sigma e^{-i\rho_\sigma(z-z_1)} & if \quad z_0 < z < z_1 \\ A^3_\sigma e^{ik_\sigma(z-z_1)} + B^3_\sigma e^{-ik_\sigma(z-z_2)} & if \quad z_1 < z < z_2 \\ A^4_\sigma e^{i\rho_\sigma(z-z_2)} + B^4_\sigma e^{-i\rho_\sigma(z-z_3)} & if \quad z_2 < z < z_3 \\ A^5_\sigma e^{ik_\sigma(z-z_3)} & if \quad z > z_3 \end{cases} \tag{3}$$

where ρ_σ denotes the wave vectors through the barriers including Dresselhaus, Rashba, and Zeeman terms are given by:

$$\rho^2_\sigma = \left(\frac{2m^*_b (V_0 - E)}{\hbar^2} \mp \frac{2\alpha_R m^*_b}{\hbar^2} k_{\parallel} \mp \frac{m^*_b g^*_b \mu^b_B}{\hbar^2} B + k^2_{\parallel} \right) \left(1 \pm \frac{2\beta_D m^*_b}{\hbar^2} k_{\parallel} \right)^{-1} \tag{4}$$

and k_σ is the reciprocal length of decay of the wave function in the wells:

$$k^2_\sigma = \frac{2m^*_w E}{\hbar^2} \pm \frac{m^*_w g^*_w \mu^w_B}{\hbar^2} B - k^2_{\parallel} \tag{5}$$

Taking into account the boundary conditions, the transmission coefficient, T_σ, can be calculated by a transfer-matrix method [15] and the spin polarization, P, that determines the difference of transparency for the spin states $|+\rangle$ and $|-\rangle$ through the EB is:

$$P = \frac{T_+ - T_-}{T_+ + T_-} \tag{6}$$

3. Results and discussions

In our numerical calculations we always chose $k_{\parallel} = 2 \times 10^8\ \mathrm{m}^{-1}$ and system electric barrier-double $a_L = a_R = 5\,\mathrm{nm}$ and $a_w = 5\,\mathrm{nm}$. The relevant parameters for non-magnetic semiconductors InAs, InSb, GaSb, and GaAs are references [11] and [16-19].

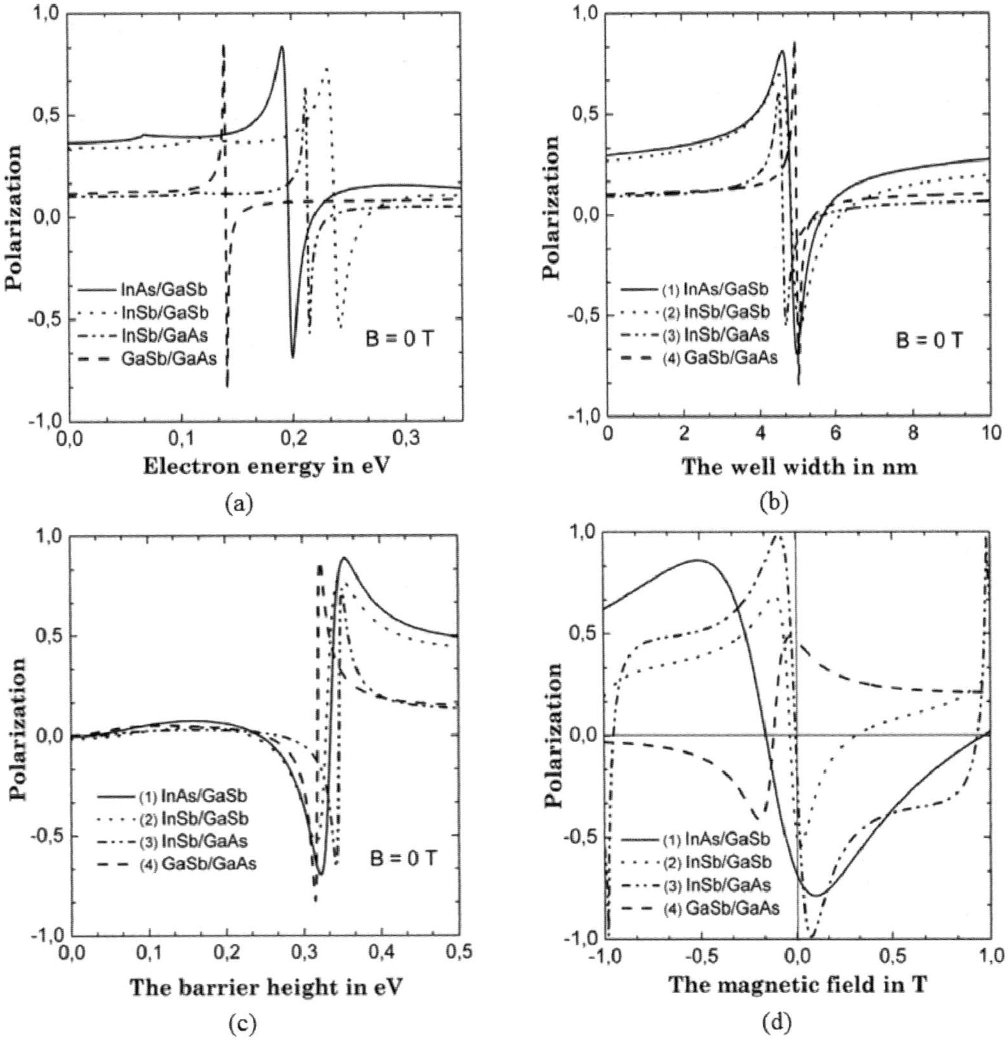

Figure 2. Symmetric double-barrier structure in the presence of the DSO and RSO for non-magnetic semiconductors: InAs/GaSb, InSb/GaSb, InSb/GaAs, and GaSb/GaAs $a_L = a_R = a_w = 5\,\text{nm}$ at fixed $k_{\square} = 2\times10^8\,\text{m}^{-1}$ (a) spin polarization as a function electron energy, the EB height $V_0 = 320\,\text{meV}$. For the following plots, the energy applied electron according to each heterostructure is $E^{(1)} = 200\,\text{meV}$, $E^{(2)} = 240\,\text{meV}$, $E^{(3)} = 220\,\text{meV}$, and $E^{(4)} = 140\,\text{meV}$. (b) Spin polarization as a function of well width. (c) Spin polarization as a function of EB height. (d) Spin polarization as a function of magnetic field with EB height $V_0 = 320\,\text{meV}$. Initial state spin with polar angle of $\varphi = \frac{1}{4}\pi$.

In Figure 2, the study is carried out at a zero magnetic field. When observing Figure 2 (a) it has the spin polarization as a function of the energy applied to the electron in heterostructures with best polarization percentages: InAs/GaSb and GaSb/GaAs with approximately 83%, but the latter with a faster relaxation spin due to the amplitude between the peaks, while the InSb/GaSb heterostructure has a lower polarization percentage, from about 67% but with a wider range of peaks generating a slower spin relaxation – ideal for manufacturing nano-devices. The figure 2 (b) shows the spin polarization as a function of width of well. We obtain resonant peaks in a width of 5 nm between for InAs/GaSb, and similar behaviors around of this value for different nanostructures. In addition, the periodicity is not seen as reported by JD Lu *et al.* [13], because in our work we report the results in the conduction band , unlike JD Lu *et al.* In Figure 2 (c) we worked spin polarization depending on the height of the EB where only one peak of about 320 meV occurs because the energy of the electrons in each non-magnetic semiconductors corresponds to resonant energy given in Figure 2 (a), without generating new periodicity.

The studying the influence of in-plane magnetic field to the non-magnetic semiconductors are shown in Figure 2 (d) two cases are presented: the first for a direction of the magnetic field $\theta = -\frac{1}{4}\pi$, that is observed on the scale of 0 to 1 T. The spin down decreases as the field is increased. Therefore the increases up spin, and the second for a direction of the magnetic field $\theta = \frac{3}{4}\pi$, that is observed on the scale of -1 to 0 T, where the phenomenon is totally opposite to the first case. Generating a spin switch of which could be developed in the process of building engineering spintronic devices

In the analysis of spin polarization in terms of the energy applied from the electron to the InAs/GaSb and InSb/GaAs heterostructures different intensities of magnetic field ($B = 0, 0.1, 1\,\mathrm{T}$) were used. Figure 3 (a) shows a resonant peak in $E = 66\,\mathrm{meV}$ and $B = 0.1\,\mathrm{T}$, where polarization is 100%, but with a deficiency in width between the very small corresponding to spin up and spin down peaks compared to those obtained at an energy of $E = 200\,\mathrm{meV}$. Here, it is also observed that for more intense magnetic fields in the order of B = 1T saturation occurs, but spin relaxation spin increases in these resonant states because the transmission coefficients T_+ and T_- are separated, as shown in Figure 4 (a).

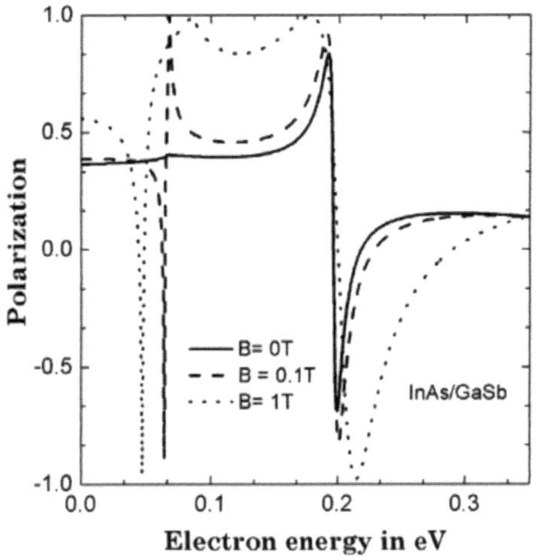

(a)

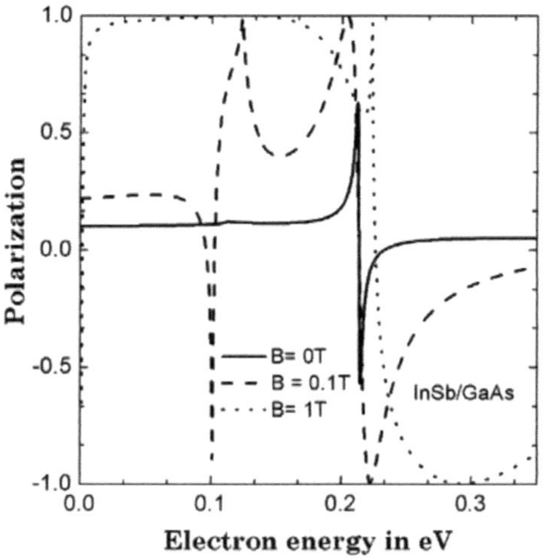

(b)

Figure 3. Symmetric double-barrier structure $a_L = a_R = a_w = 5\,\mathrm{nm}$ at fixed $k_{\parallel} = 2\times10^8\,\mathrm{m}^{-1}$, and EB height $V_0 = 320\,\mathrm{meV}$. Spin polarization in the presence of the Dresselhaus and Rashba spin-orbit coupling as a function of electron energy to different magnetic field for the heterostructure: InAs/GaSb and InSb/GaAs. Energy applied electron is $E = 200\,\mathrm{meV}$ for (a) and $E = 220\,\mathrm{meV}$ for (b). Initial state spin of $\varphi = \frac{1}{4}\pi$.

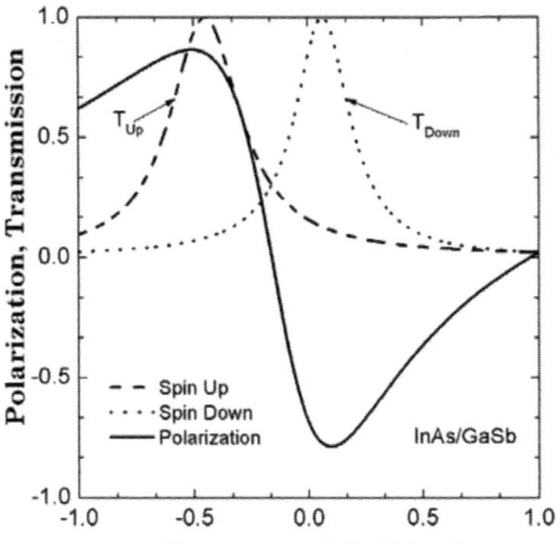

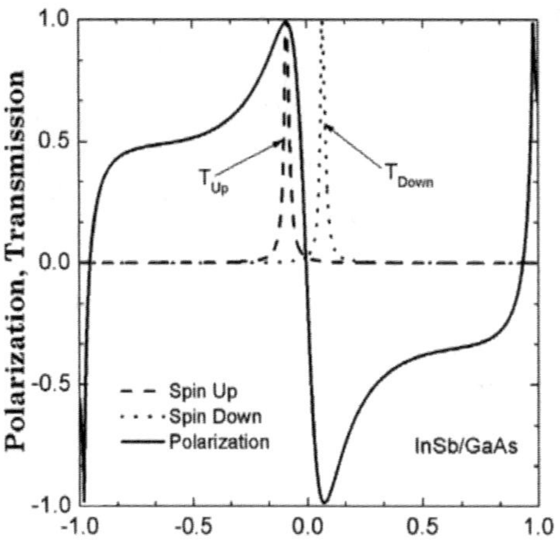

Figure 4. Symmetric double-barrier structure $a_L = a_R = a_w = 5\,\text{nm}$ at fixed $k_{\parallel} = 2\times10^8\ \text{m}^{-1}$, and EB height $V_0 = 320\,\text{meV}$. Spin polarization and transmission coefficient in the presence of the Dresselhaus and Rashba spin-orbit coupling as a function of magnetic field for the heterostructure InAs/GaSb and InSb/GaAs. Energy applied electron is $E = 200\,\text{meV}$ for (a) and $E = 220\,\text{meV}$ for (b). Initial state spin with polar angle of $\varphi = \frac{1}{4}\pi$.

In heterostructures shown in Figures 3 and 4 (b) a high sensitivity to the effects of low magnetic field is observed, reaching 100% polarization. With a spin relaxation (change of spin down to up) favorable since the separation between the peaks is not abrupt; suggesting future studies of spintronics in asymmetric double barrier heterostructures of InAs/GaSb/InAs/GaAs/InAs types.

4. Conclusions

We have theoretically investigated electron transport through spin-dependent non-magnetic semiconductors with DSO and RSO. When the heterostructure does not include magnetic field, it only has the spin-orbit coupling; we predict that spin-polarization spontaneously emerges in this heterostructure due to the combination of the coherence interaction effect of the DSO and RSO. But if InAs / GaSb and InSb / GaAs heterostructures are influenced by the magnetic field due efficiency of spin polarization and the probability transmission obtained in this work; that is very important for purposes of application, since the energy scales applied to electron and magnetic field is experimentally controllable, as the field direction.. Therefore, spin relaxation can be thought out that depends strongly on this compensation DSO and RSO coupling [20], which can be used in fabrication of spin dependent optoelectronic devices.

By studying the influence of in-plane magnetic field in the $\theta = -\frac{1}{4}\pi$ direction, which decreases the spin down to measure magnetic field increases, therefore, generating spin up increases, totally opposite case when the magnetic field is oriented to $\theta = \frac{3}{4}\pi$, generating a spin switch that could be developed during the process of engineering building spintronic devices.

5. Acknowledgments

The authors thank the vice-rectories for research (VRI) and administration at Universidad del Cauca for partial funding to develop this work through research projects: ID 3088 and ID 4027.

References

[1] Wolf S A, Awschalom D D, Buhrman R A, Daughton J M, Von Molnar S, Roukes M L, Chtchelkanova A Y and Treger D M 2001 *Science* **294** 1488
[2] Zuti´c I, Fabian J and Das Sarma S 2004 *Rev. Mod. Phys.* **76** 323

[3] Bader S D and Parkin S S 2010 *Annu. Rev. Condens. Matter Phys.* **1** 71
[4] Nielsen M A and Chuang I L 2011 *Quantum Computation and Quantum Information* (Cambridge: Cambridge University Press)
[5] Koga T, Nitta J, Takayanagi H, Datta S 2002 *Phys. Rev. Lett.* **88** 126601
[6] Wang L G, Yang W, Chang K 2005 *Phys. Rev. B* **72** 153314
[7] Ye C Z , Zhang C X, Nie Y H and Liang J Q 2007 *Phys. Rev. B* **76** 035345
[8] Li W and Guo Y 2006 *Phys. Rev. B* **73** 205311
[9] Papp G and Peeters F M 2001 *Appl. Phys. Lett.* **78** 2184
[10] Papp G and Peeters F M 2001 *Appl. Phys. Lett.* **79** 3198
[11] Lu J D and Li Y B 2010 *Superlattices Microstruct* **48** 517
[12] Glazov M M, Alekseev P S, Odnoblyudov M A, Chistyakov V M, Tarasenko S A, Yassievich I N 2005 *Phys. Rev. B* **71** 155313
[13] Lu J D, Xia X P, Yi L and Wang Y H 2010 *Phys. Lett. A* **374** 3341
[14] Sun Q F and Xie X C 2006 *Phys. Rev. B* **73** 235301
[15] Ferry D K and Gooodnick S M 1997 *Transport in Nanostructures* (UK: Cambridge University Press) p 108-111
[16] Gnanasekar K and Navaneethakrishnan K 2005 *Physica E.* **28** 328
[17] Lu J D and Li I W 2005 *Appl. Surface Sciences* **256** 4027
[18] Peter A J 2008 *Phys. Let. A.* **372** 5239
[19] Toloza Sandoval M A, Ferreira da Silva A and La Rocca G C 2012 *Phys. Procedia* **28** 95
[20] Khayatzadeh Mahani M R 2008 *Phys. Lett. A.* **372** 6022

Preparation of Fe_3O_4 Nanoparticles and Removal of Methylene Blue through Adsorption

J. S. Trujillo Hernandez[1], A. Aragón Muriel[2], J. A. Tabares[1], G. A. Pérez Alcázar[1] and A. Bolaños[2]
[1] Departamento de Física, Universidad del Valle, Meléndez, A.A. 25360, Cali, Colombia.
[2] Departamento de Química, Universidad del Valle Meléndez, A.A. 25360, Cali, Colombia.

E-mail: juan.sebastian.trujillo@correounivalle.edu.co

Abstract. In this work, we studied the catalytic activity, structural properties, and behavior of a Fe_3O_4 magnetic system. The Fe_3O_4 nanoparticles were prepared by the thermal decomposition method. X-ray diffraction confirmed the presence of a structural Fe_3O_4 phase, where acicular shape of the grains is shown. Transmission Mössbauer spectroscopy showed a wide distribution of particle sizes at room temperature, some of these present superparamagnetic behavior and are responsible of paramagnetic sites. The hysteresis loops obtained by the use of a vibrating sample magnetometer showed that these nanoparticles exhibit superparamagnetic behavior. However, the cycles present a significant contribution from a ferrimagnetic component at 2 K, which agrees with Mössbauer results. Through scanning electron microscopy, a tendency to the agglomeration of nanoparticles was observed. Nanoparticle activity in the degradation of methylene blue (MB) was studied through fluorescence spectroscopy, finding dye adsorption properties.

1. Introduction

The magnetic properties of iron oxide nanoparticles have been widely applied in the electronics industry [1-2]; however, currently investigations are conducted in the search for pharmaceutical [3] and catalytic applications [4-6]. These magnetic properties, coupled with the surface area of the nanoparticles of magnetite, are among the major advantages at the time of studying catalytic activity, providing new analytical chemistry methods for removing metals in water [7-8] and contaminants dyes degrade in aqueous matrices [9-10]. Methylene blue dye is widely used in staining procedures; to examine DNA or RNA in microscopy studies, and as redox indicator. Despite its health benefits, in higher doses methylene blue can take electrons from complexes of the electron transport chain, becoming a health risk. In many catalytic applications, Fe3O4 nanoparticles are used as a metal complex support material or to be functionalized with other materials (gold, manganese complexes, polyethyleneimine), which enhances activity [11-13], so it is important to carry out studies with magnetite before functionalization.

In this paper, magnetite nanoparticles were obtained from ferric solutions. During characterization, a particle size distribution was observed, where some showed superparamagnetic behavior. Nanoparticle activity in the degradation over time of methylene blue was studied via fluorescence spectroscopy.

2. Experimental procedures

2.1 Synthesis of Fe_3O_4 nanoparticles

Synthesis of magnetite nanoparticles was performed as reported in the literature [14]. Iron chloride tetrahydrate [$FeCl_2{\cdot}4H_2O$] (278.3 mg, 1.4 mmol); and sodium oleate (1851.2 mg, 5.1 mmol) were put into two-neck round bottom flask were stirred in a mixture (280 ml) of ethanol:water:hexane (4:3:7). The resulting solution was heated to 70 °C and kept at that temperature for 4 h. When the reaction was completed, the upper organic layer containing Fe_3O_4-oleate complex was washed three times with 30 ml distilled water. After washing, hexane was evaporated-off, the resulting Fe_3O_4-oleate complex was introduced into a pyrex tube and then sealed less than 40 Pa. The pyrex tube was heated to 380 °C for 4 h. The reaction mixture was cooled to room temperature, yielding a black fluid. The Fe_3O_4 nanoparticles produced dispersed in chloroform. To precipitate the particles, 20 ml of ethanol was added. The particles were isolated by centrifuge, and then dissolved in 20 ml of hexane. After adding 0.02 ml of dispersion agent, the precipitate appeared and was isolated by centrifuge. The precipitate was washed with ethanol and centrifuged to get the black solid product, which was washed again with ethanol, centrifuged, and finally re-dispersed in hexane and stored under argon.

2.2. Characterization

The sample prepared was characterized by the following techniques: X-ray diffraction, Mössbauer spectroscopy, vibrating sample magnetometer, scaning electron microscopy, and fluorescence spectrometer. The X-ray measurements were obtained at room temperature for the sample by using a PANalytical diffractometer equipped with a Cu target X-ray tube and a Ni filter to select the Cu-$K\alpha$ radiation. The X-ray profile was collected from 20-90° in scattering angle using a 0.02° step and counting times of 2 s per point. The patterns were refined by the Rietveld method combined with Fourier analysis to describe the broadening of the lines by using the GSAS program [15].This yielded the average values of the lattice parameter, crystallite size and the structural phase. Mössbauer measurement was performed at room temperature with transmission geometry using a 57Co(Rh) source of 25 mCi. The spectra were fitted with the MOSFIT program [16], taking an α-Fe foil as the calibration sample and using a HMFD and two paramagnetic sites. The behavior of magnetization as a function of the field was studied by using VSM and using a physical properties measurement system (PPMS) available a the Center of Excellence for Novel Materials (CENM). Surface morphology of the sample was observed by using SEM.

2.3. Adsorption studies

The degradation activity of nanoparticles was assessed for cationic (MB) dye. The temperature and shaking speed for all the adsorption experiments were kept constant at 25.0 ± 0.1 °C and 130 rpm, respectively. Typically 10 ml of dye solution with concentration of 20 $mg{\cdot}L^{-1}$ MB was mechanically shaken with 0.2 $g{\cdot}L^{-1}$ Fe_3O_4. The concentrations of residual MB, after magnetic separation of Fe_3O_4, were measured with the emission band at 680 nm (λ_{exc} = 633 nm) using a fluorescence spectrometer (FP-8500, JASCO). The excitation lambda (633 nm) corresponds to the most intense band after performing the sweep from the UV region at 230 nm to the UV-Vis region at 800 nm.

3. Results and Discussion

3.1 *X-ray diffraction, morphology and size distribution results*

Figure 1 shows XRD patterns indicating the position and intensities corresponding to the spinel structure of the (Fe3O4) magnetite peaks (JCPDS 01-11111 data) [10, 17]. From the refinement of the diffractogram of Fe3O4 nanoparticles it was obtained that: crystallite size parallel and perpendicular equal to 79.5 ± 1 Å and 137.2 ± 2 Å, respectively, indicating that the crystallites have acicular shape; and lattice parameters of this structure are: a, b, c = 8.379 Å. According to the micrograph obtained via SEM, and shown in Figure 2, it can be noted that the particles present an agglomeration tendency. Z potential measurements showed particle sizes between 15 and 100 nm, with an average 42 nm particle size.

3.2. Mössbauer and magnetic results

The Mössbauer spectrum (Figure 3) was fitted with two sextets, one doublet, one singlet and a hyperfine field distribution due to the distribution of particle sizes [17]. This is caused by the presence of various types of crystallographic environment of iron atoms (Fe^{2+} and Fe^{3+}) into the spinel structure of the magnetite, where the Fe^{2+} and Fe^{3+} ions are located in different crystallographic positions (octahedral and tetrahedral) [17]. The ionic state of the Fe^{2+} and Fe^{3+} atoms can not be distinguished at room temperature as a result of fast electron hopping [17]; this suggests that the presence of components with lower hyperfine field is due to reduced particle size. The central part of the spectrum is a consequence of particle size distribution; this suggests that refined particle sizes gives a superparamagnetic behavior and large particle sizes gives ferrimagnetic behavior. Table 1 shows the Mössbauer parameters obtained. Trough magnetization it was observed that the Fe_3O_4 magnetite sample at room temperature presents a superparamagnetic behavior and at low temperatures 2 K it shows a widening of the hysteresis loop, presenting ferrimagnetic behavior with coercive field of 613.2 Oe and a saturation magnetization of 50.3 emu/g (Figure 4). The low value of magnetization with respect to the value of the bulk (92 emu/g) is probably due to the mass of the surfactant (sodium oleate), given the existence of a surface layer of the surfactant in the particles decreases the value of the magnetization because this can make a diamagnetic contribution [17]. The different behaviors shown by Mossbauer (superparamagnetic-ferrimagnetic) and hysteresis loops (superparamagnetic) are a consequence of the different measurement times of both techniques.

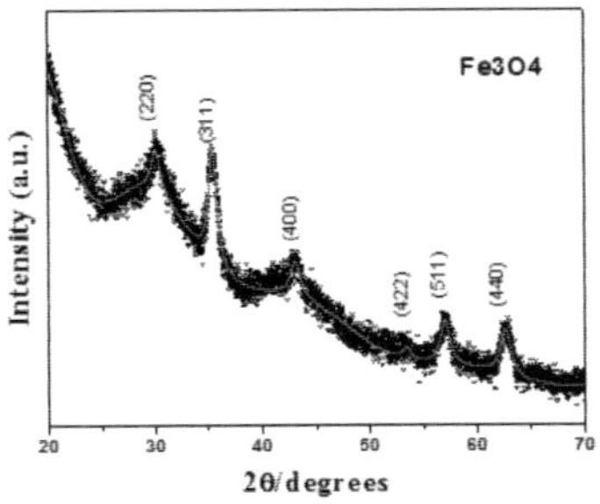

Figure 1. XRD patterns of Fe_3O_4

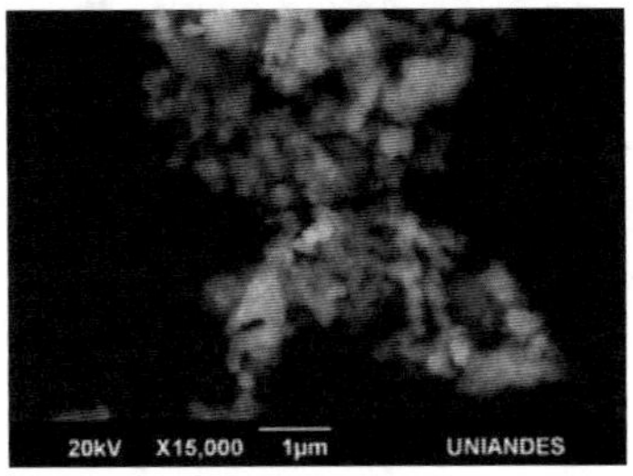

Figure 2. SEM micrographs of Fe_3O_4.

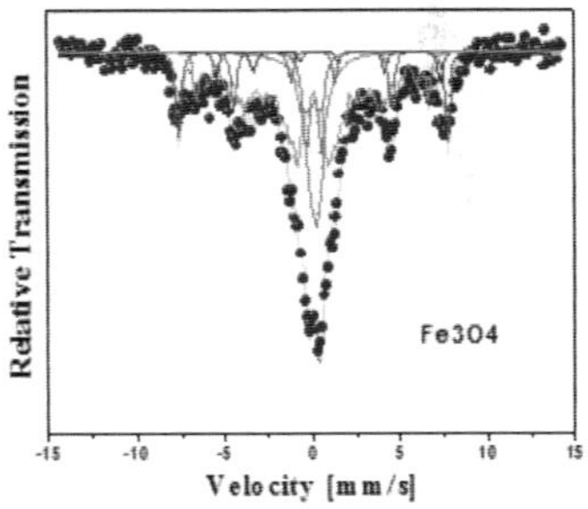

Figure 3. Mössbauer of Fe_3O_4

1.3. Adsorption studies

Adsorptive removal activity of the sample prepared was evaluated by measuring the degradation rate of MB in the presence of magnetite nanoparticles. It was found that the MB concentration remained stable after 1 h of stirring in the presence of the powder obtained. The results of MB degradation at different adsorption times of Fe_3O_4 are shown in Figure 5. Emission spectra obtained after exciting at 633 nm; showed a decrease in the intensity of the emission band observed at 680 nm, indicating that the MB concentration decreases with nanoparticle adsorption, finding particular properties in the degradation of this cationic dye. The activity may increase if photocatalytic studies are conducted by using UV irradiation in the presence of the nanoparticles, as has been done in similar studies [9-10].

Conclusions

In this study, magnetite particles with an average 42 nm diameter were prepared by thermal decomposition. According to magnetization and Mössbauer spectroscopy data for Fe_3O_4 particles, this indicates that the particles have superparamagnetic properties that disappear when the temperature drops. The transformation has been observed to the ferrimagnetic state in the range of 2 K, which can be explained by increased dipole-dipole interactions between the magnetic particles. The Mössbauer spectroscopy data for magnetite particles do not show a stable superparamagnetic fraction, and hyperfine magnetic structure appeared in the absence of H_{ext}. The study of the magnetite activity in the degradation of methylene blue shows that the nanoparticles obtained have adsorbent properties for the cationic dye over time; this is evidenced by a decrease in intensity of the emission band obtained by fluorescence. This study could be applied in wastewater decontamination.

Acknowledgements: This work was supported by CENM (Center of Excellence for Novel Materials).

Table 1. Hyperfine parameter: isomer shift (IS), mean hyperfine field <B_{hf}>, quadrupole splitting (QS), line width (GA) and percentage of area of the Fe_3O_4 nanoparticles.

	Hyperfine Parameter				
Component	<B_{hf}> ± 0.2 [T]	QS ± 0.05 [mm/s]	IS ± 0.05 [mm/s]	GA ± 0.05 [mm/s]	Area [%]
Sextet 1	39.2	0.50	0.41	0.30	4.88
Sextet 2	47.4	-0.01	-0.23	0.30	13.14
Distribution	39.1	-0.01	-0.22	0.30	52.37
Doublet	---	0.70	-0.19	0.40	9.08
Singlet	---	0.00	-0.13	1.00	20.53

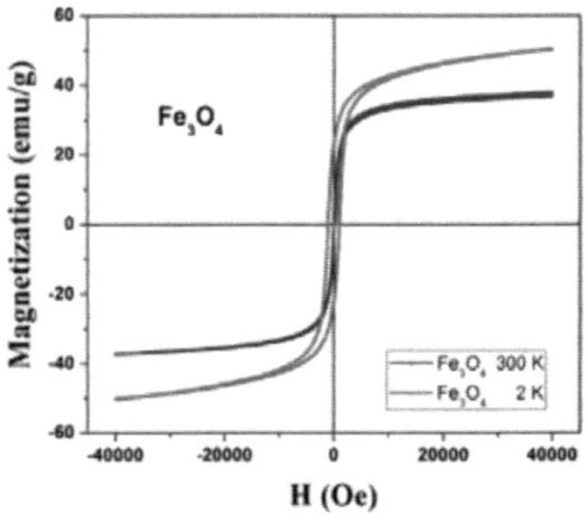

Figure 4. Hysteresis loop of Fe_3O_4

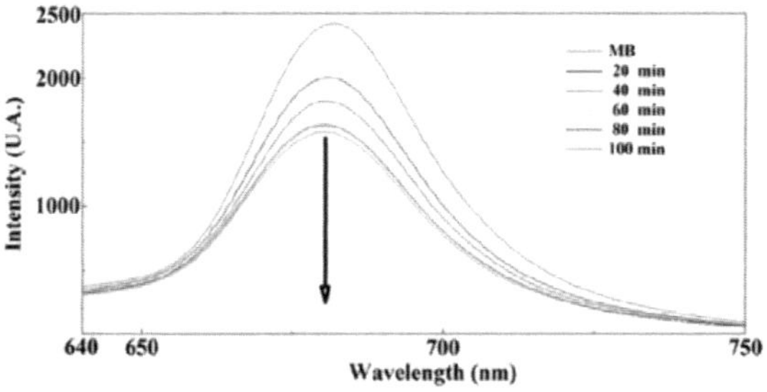

Figure 5. Emission spectra of MB vs. degradation time in the presence of Fe_3O_4

References

Younghoon K., Kyungyun K., Sun Kak H., Cheolmin P., Jinhan C.: *ACS Nano*, 2014, **8**, 2419-2430.

Siva C., Karthikeyan B., Solomon J. S., Baraneedaran P., Annie R. S., Sivakumar M.: *Mater. Sci. Forum*, 2014, **781**, 1-16.

Mesárošová M., Kozics K., Bábelová A., Regendová E., Pastorek M., Vnuková D., Buliaková B., Rázga F., Gábelová A.: *Toxicol. Lett.*, 2014, **226**, 303–313.

Gawande M. B., Branco P. S., Varma R. S.: Chem. Soc. Rev., 2013, **42**, 3371-3393.

Govan J., Gun'ko Y. K.: *Nanomater.*, 2014, **4**, 222-241.

Shahbazi F., Amani K.: *Catal. Commun.*, 2014, **55**, 57–64.

Giraldo, L., Erto A., Moreno-Pirajan J.C.: *Adsorpt.*, 2013, **19**, 465–474.

Mahdavi S., Jalali M., Afkhami A.: *J. Nanopart. Res.*, 2012, **14**, 171.

Hong-Jun T., Ting-Ting H., Zhi-Jun L., Xiang-Yang W.: *Chin. Chem. Lett.*, 2013, **24**, 63–66.

Giri S. K., Das N. N., Pradhan G.C.: *Colloids and Surfaces A: Physicochem. Eng. Aspects*, 2011, **389**, 43-49.

Pariti A., Desai P., Maddirala S. K. Y., Ercal N., Katti K. V., Liang X. Nath, M.: *Mater. Res. Express*, 2014, **1**, 1-19.

Bagherzadeha M., Haghdoosta M. M., Moghaddama F. M., Foroushania B. K., Saryazdia S., Payaba E.: *J. Coord. Chem.*, 2013, **66**, 3025-3036.

Cai H., An X., Cui J., Li J., Wen S., Li K., Shen M., Zheng L., Zhang G., Shi X.: *Appl. Mater. Interfaces*, 2013, **5**, 1722–1731.

Bronstein L. M., Huang X., Retrum J., Schmucker A., Pink M. , Stein B. D., Dragnea B.: *Chem. Mater.*, 2007, ***19***, 3624-3632.

Larson A. C., Von Dreele R. B.: General Structure Analysis System (GSAS), Los Alamos National Laboratory Report LAUR, 1994, 86–748.

Teillet J., Varret F.: Mosfit Programm, University du Maine, unpublished, 1976.

Dutta R. K., Sahu S., Reddy V. R.: *J. Nanopart. Res.*, 2012, **14**, 1052.

Effect of annealing treatment on the photocatalytic activity of TiO_2 thin films deposited by dc reactive magnetron sputtering

L.M. Franco Arias[1], A. Arias Duran[1], D. Cardona[2], E. Camps[2], M.E. Gómez[1], G. Zambrano[1]*

[1] Thin Film Group, Department of Physics, Universidad del Valle, Calle 13 # 100-00, Cali, Colombia

[2] Department of Physics, ININ, Apartado Postal 18-1027 México DF 11801, México

E-mail: gustavo.zambrano@correounivalle.edu.co

Abstract. Titanium dioxide (TiO_2) thin films have been deposited by DC reactive magnetron sputtering on silicon and quartz substrates with different Ar/O_2 ratios in the gas mixture. Substrate temperature was kept constant at 400 °C during the deposition process, and the TiO_2 thin films were later annealed at 700 °C for 3 h. The effect of the Ar/O_2 ratio in the gas flow and the annealing treatment on the phase composition, deposition rate, crystallinity, surface morphology and the resulting photocatalytic properties were investigated. For photocatalytic measurements, the variation of the concentration of the methylene blue (MB) dye under UV irradiation was followed by a change in the intensity of the characteristic MB band in the UV–Vis transmittance spectra. We report here that the as-grown TiO_2 films showed only the anatase phase, whereas after annealing, the samples exhibited both the anatase and rutile phases in proportions that varied with the Ar/O_2 ratio in the mixture of gases used during growth. In particular, the annealed TiO_2 thin film deposited at a 50/50 ratio of Ar/O_2, composed of both anatase (80%) and rutile phases (20%), exhibited the highest photocatalytic activity (30% of MB degradation) compared with the samples without annealing and composed of only the anatase phase.

1. Introduction

Titanium dioxide is a material known for its high refractive index, high transparency in the visible and near-infrared wavelength region [1], high dielectric constant, very good wear resistance and stability and a large semiconductor band gap, which has developed significant importance as a catalyst for the breakdown of organic substances under UV illumination. Due to these special properties, TiO_2 has become the subject of much research for applications in optical coatings, microelectronic devices, and protective layers [2]. Crystalline TiO_2 films are a focus of interest because of their specific properties compared to the amorphous phase. Titanium dioxide crystallizes in three structures: tetrahedral anatase, rutile and orthorhombic brookite. Rutile is primarily desirable for optical applications and is the thermodynamically stable phase; anatase has high kinetic stability and is more hydrophilic and photocalytic [3], which allows the creation of products with new properties such as easy-to-clean surfaces, self-cleaning windows, as well as photocatalytic air and water purification devices. However, it has been argued that a mixture of anatase and rutile may show the highest photocatalytic activity; an example is the P25 powder from Degussa [4]. Many researchers [5–7] have investigated mixtures of the TiO_2 phases regarding their performance as photocatalysts. In this work, we deposited TiO_2 thin films by DC magnetron sputtering at 400 °C onto quartz substrates and studied the crystallinity, phase mixing, and photocatalytic properties, such as the dependence on the Ar/O_2 ratio in the gas mixture and the subsequent annealing treatment.

2. Experimental details

We grew TiO_2 films on quartz and silicon (100) substrates with a DC reactive magnetron sputtering system from a titanium target (99.9%) with 2.54 cm diameter that was placed inside a chamber with

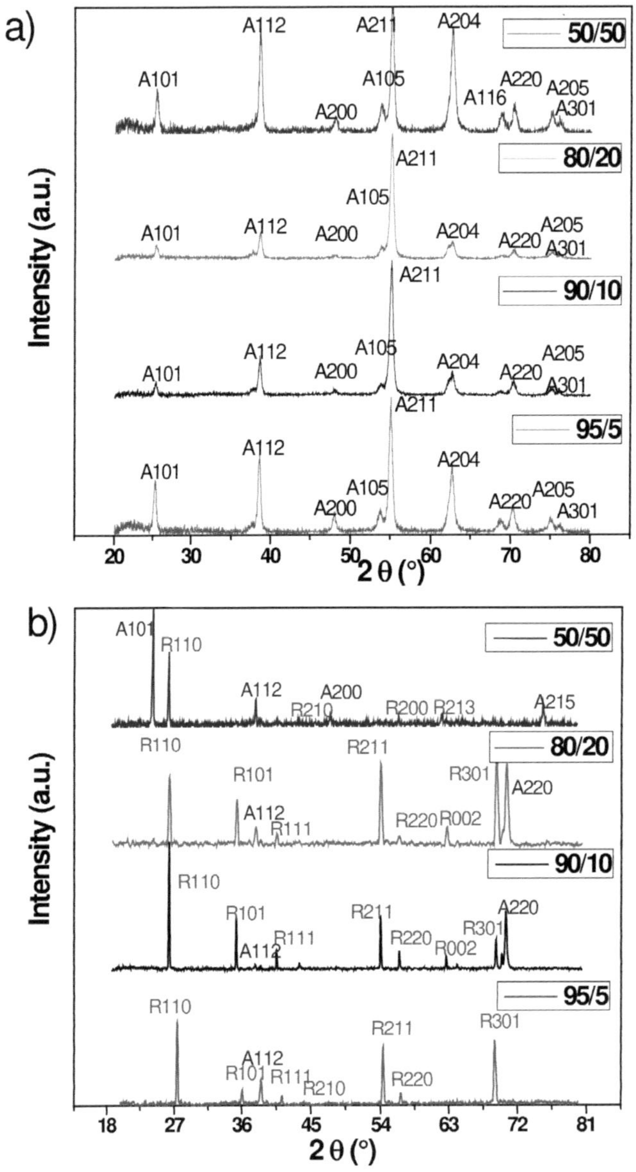

Figure 1. TiO_2 XRD patterns of films deposited on quartz at different Ar/O_2 flow ratios: a) unannealed and b) annealed

an Ar/O_2 gas mixture. The Ar/O_2 ratio flow in the gas mixture was systematically varied (95/5, 90/10, 80/20, and 50/50) to study the influence of the oxygen partial pressure on the crystallographic structure and photocatalytic activity of the TiO_2 thin films. The base pressure was 4 x 10^{-4} mbar, and the working pressure (2.4 x 10^{-2} mbar), power applied to the target (150 W), distance from the substrate to the Ti target (8 cm), temperature (400 °C), and deposition time (2 h) were maintained constant for all samples. After the sputtering process, the TiO_2 films were annealed at 700 °C for 3 h. The crystal structure of the TiO_2 films was determined by using glancing angle X-ray diffraction (GAXRD) at a 2° incidence angle in the 2-theta mode over the range of $20° < 2\theta < 80°$ with a Bruker

(Model D8 Advance) diffractometer using the Kα1 line of Cu (λ = 1.5406 Å, 25 kV and 16 mA). The crystallite size was deduced from the full width at half maximum of the XRD lines according to Scherrer´s equation [8]. Raman spectra were taken on a JobinYvon micro-Raman system. For the photocatalytic measurements, the variation of the concentration of the methylene blue (MB) dye was followed by the change (a decrease) of the characteristic band intensity of the MB, located at 665 nm, in the UV–Vis transmittance spectra, which was analyzed as a function of the UV irradiation time. The MB solution in a quartz cell was irradiated with ultraviolet light with a power of 4 W and the source was located 4 cm from the cell. The solution was bubbled with air before irradiation. From these measurements, we evaluated photocatalytic performance by analyzing the percentage of dye degradation when the films were used as a catalyst in a Shimadzu spectrophotometer (UV model 1601).

3. Results and discussion

3.1 XRD analysis

The XRD patterns of TiO_2 films deposited on quartz substrates at ratios of 95/5, 90/10, 80/20, and 50/50 of Ar/O_2 are displayed in Figure 1a). We can observe that all of the TiO_2 films show a preferred (211) orientation (denoted as A211) of the anatase crystalline structure; however, the more commonly reported anatase phase orientation is the (101) orientation. This behavior can be attributed to the fact that different surface free energies are associated with the different diffraction planes [9]. In our case, the applied sputtering power density (~300 kWm^{-2}**) activates TiO_2 film growth from a relatively lower surface free energy to a higher surface free energy [10]. The crystallite size of the samples calculated from the Scherrer equation were 18.6, 18.6, 17.9, and 17.7 ± 0.1 nm for the films deposited at ratios of 95/5, 90/10, 80/20 and 50/50 of Ar/O_2, respectively. The XRD patterns of the films subjected to heat treatment at 700 °C for 3 h are shown in Figure 1b). In the patterns, increased film crystallinity with annealing and change to the anatase-rutile phase mixture can be observed. The films deposited at 80/20 and 90/10 crystallized in completely different manner, with peaks predominately in the rutile phase with preferred orientations in the 301 and 110 planes (peaks noted as R301 and R110) and less appearance of the anatase peaks. From Figure 1b), it is clear that the structure of the films after annealing is strongly influenced by the Ar/O_2 ratio, presenting a tendency

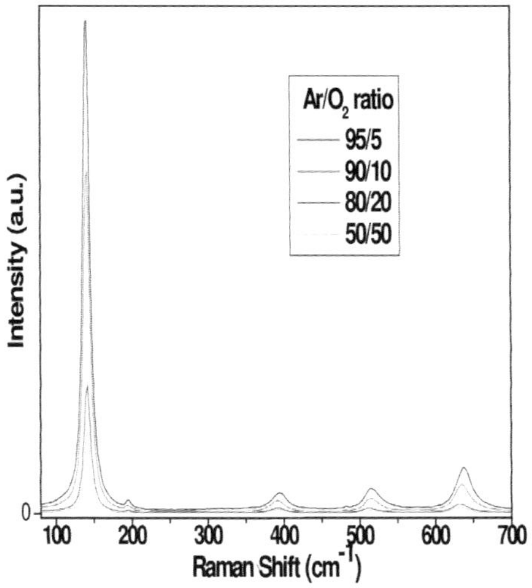

Figure 2. Raman spectra of unannealed TiO_2 films deposited at different Ar/O_2 flow ratios

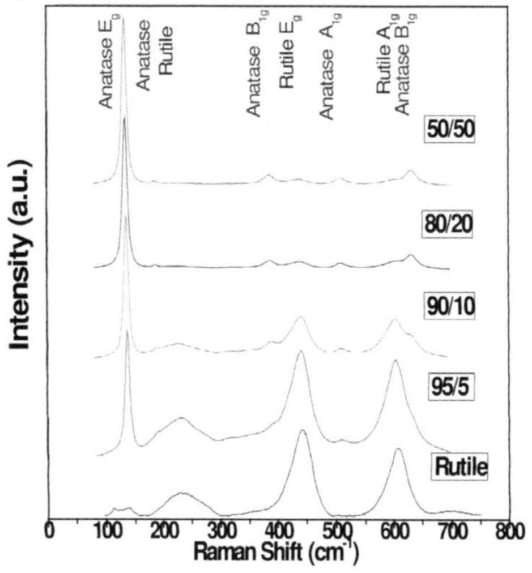

Figure 3. Raman spectra of TiO_2 films annealed for 3 h at 700 °C at different Ar/O_2 flow ratios

for the rutile phase to increase when the Ar/O_2 ratio decreases. We can conclude that for unannealed TiO_2 films, the crystallite size of the anatase phase remains almost constant (~18 nm) with increasing oxygen percentage, whereas for the annealed films, the crystallite size of the anatase and rutile phases decreases from 38.9 to 28.2 and from 49.4 to 31.7 nm, respectively; in both cases, it is higher than the crystallite size in the unannealed TiO_2 films.

Table 1. Percentages of anatase and rutile phases for the annealed samples, calculated from the Raman analysis

Ar/O_2 ratio	Anatase (%)	Rutile (%)
95/5	29	71
90/10	42	58
80/20	65	35
50/50	80	20

Table 2. Photocatalytic degradation of MB in presence of annealed TiO_2 films

Ar/O_2 ratio	Degradation %	Anatase/Rutile relation		Crystallite size (nm)	
		Rutile %	Anatase %	Rutile	Anatase
95/5	13	71	29	49.4	38.9
90/10	25	58	42	47.3	33.3
80/20	29	35	65	36.6	29.4
50/50	30	20	80	31.7	28.2

3.2 Raman analysis

To study how the Ar/O_2 flow ratio affects both the anatase and rutile phase formation in the TiO_2 films, we measured the Raman spectra of the films deposited on quartz substrates without annealing. Figure 2 displays these spectra. All spectra show the presence of an intense peak at the 142.2 cm^{-1} wavenumber associated with the TiO_2 E_g band of anatase, as previously reported [11]. Similarly, all of the spectra display three smaller peaks at 394.9 (associated with the B_{1g} mode of the anatase phase), 515.5 and 637.4 cm^{-1}, which agree with those reported for the structure of the anatase phase [10]. Moreover, the intensity for each peak increases, and the peaks become narrower with decreasing oxygen partial pressure in the Ar/O_2 gas mixture. This behavior indicates that the samples organize into microstructures with better crystallinity as the partial pressure of oxygen decreases [12]. Considering the phase diagram proposed by Zeman *et al.* [13], the TiO_2 film deposited at 2.4 Pa working pressure and at a smaller oxygen amount in Ar/O_2 ratio (95/5) has an anatase phase crystallization higher than those of the other samples, as shown in Figure 2. To induce the appearance of the rutile phase in samples reaching the anatase phase, we subjected them to annealing at 700 °C for 3 h. Figure 3 shows the Raman spectra after heat treatment. We can clearly observe that the anatase phase is transformed into the rutile phase in different proportions depending on the as-grown Ar/O_2 ratio flow. The percentages of the corresponding phases were obtained by performing the decomposition of the anatase and rutile peaks for each sample. Decomposition was performed by calculating the percentage of each of the areas with a combination of Lorentzian and Gaussian fits. Table 1 presents the corresponding calculated percentages of the phases. According to the phase diagram presented by Löbl *et al.* [14], the samples with a higher degree of crystallization in the anatase phase are expected to be transformed more easily to the rutile phase after the annealing process. Indeed, under our conditions, this effect occurred for the samples deposited at the 95/5 ratio of Ar/O_2, as observed in Figure 3 and Table 1.

3.3 Photocatalytic activity evaluation

Evaluation of the photocatalytic activity was initially carried out by irradiating the TiO_2 samples with ultraviolet light. Figure 4a) displays the relative concentration reduction of MB as a function of the UV irradiation time for unannealed TiO_2 films deposited at different Ar/O_2 ratios. We observed the degradation of the MB compound. The percentage of relative degradation of MB in the unannealed films was approximately 26%, considering that all unannealed samples showed only the anatase phase and had almost the same gap value [15] and a very similar crystallite size. The

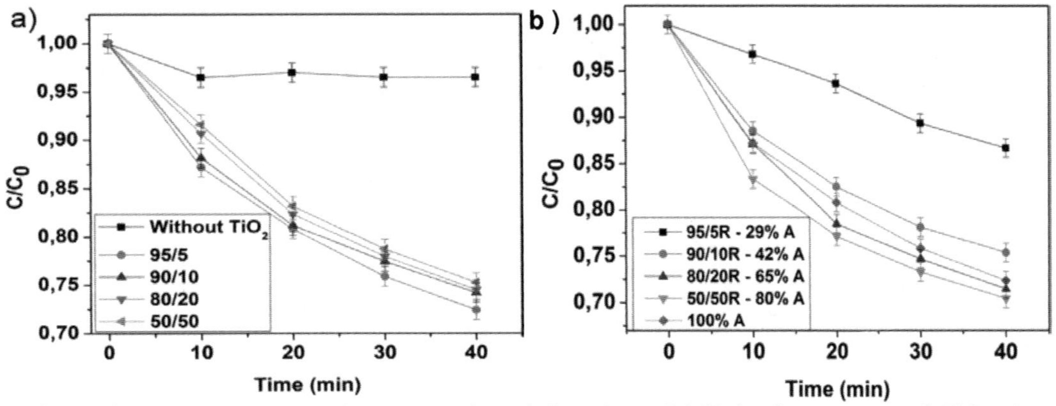

Figure 4. Relative concentration versus degradation time of MB in the presence of TiO_2 a) unannealed and b) annealed films deposited at different Ar/O_2 ratios

anatase phase has a band gap of 3.2 eV and an affinity adsorption of organic compounds much higher than the rutile phase [16]. Additionally, the anatase phase has a lower recombination rate than the rutile phase because its hollow capture rate is 10 times higher [17]. In contrast, the thermodynamically stable rutile phase has a smaller gap of 3.0 eV, which allows excitation by wavelengths extending into the visible range at 410 nm. However, the anatase phase of TiO_2 is generally considered the most active phase of photochemistry, most likely due to the combined effect of a lower recombination rate and higher adsorption capacity on the surface. Other research has investigated TiO_2 anatase and rutile phases mixing and found that these have a higher photocatalytic activity than does either individual pure phase [18]. Therefore, because our TiO_2 samples initially presented only the anatase phase, the appearance of the rutile phase was induced by annealing at 700 °C for 3 h. Figure 4b) shows the results of photocatalysis tests of the annealed samples. The percentage of relative degradation of MB varies from 13% for the 90/5 grown film with 71% rutile and 29% anatase relative composition up to 30% for the 50/50 with 20% rutile and 80% anatase relative composition.

The results are shown in Table 2. Based on reports in several papers [20-22], the rutile phase has low photocatalytic activity; therefore, no test samples have been made on a 100% rutile phase sample. As mentioned above, the two polymorphs of the TiO_2 photocatalysts in the mixed phase can effectively reduce the recombination of photogenerated electron-hole pairs to improve photocatalytic activity. In addition, photoactivity depends on crystallite size and, as noted above, the region where the potential barrier is generated should be in the order of the crystallite size. If the crystallite is very small, the carriers easily reach the surface to participate in photoactivity. In large crystallites, the separated charges undergo recombination outside the barrier region. Therefore, the probability of recombination is high, as reflected in the photoactivity. The radiation absorbed outside the barrier region behaves similarly to the pure anatase and rutile phases, in a manner unaffected by the effects of the interface. Samples containing 65and 80% of the anatase phase have higher photoactivity than do samples containing only the pure anatase phase. According to Tanaka *et al.* [19], the photocatalytic activity of TiO_2 is determined by the anatase phase content and crystallite size. These authors reported [19] a proportional increase in photocatalytic activity with increased TiO_2 crystallite size. However, in our case by comparing the crystallite sizes for both phases (Table 2), we can conclude that decreased crystallite size increases photocatalytic activity. The higher photocatalytic activity occurs in the sample with a rutile-anatase ratio of 80/20, and this photocatalytic activity is similar to the commercial TiO_2 powder Degusa 25 [23, 24] that is widely used for photocatalysis. We can also conclude that increasing the content of the anatase phase in the sample increases the photocatalytic activity. Regarding the rutile phase, this result may be related to the increased affinity of absorption of organic compounds with this phase.

Conclusions

We found that the oxygen content in the Ar/O_2 ratio in the gas mixture during growth using a sputtering technique strongly influences the deposition rate, the structure of the films after annealing, and the photocatalytic activity. After thermal treatment, higher oxygen content in the Ar/O_2 ratio improves the crystallinity of the as-grown TiO_2 films, and the samples exhibit a tendency to decrease their rutile phase content. We also found that the phase mixture improves photocatalytic activity compared to samples exhibiting only the anatase phase. The best percentage of MB degradation (30%) was for the sample with an anatase to rutile ratio of 80/20.

Acknowledgments

This work was supported by funds for internal calls for projects at Universidad del Valle 2012 IC7899 and the Center of Excellence for Novel Materials (CENM).

References

[1] Narashimha Rao K, Mohan S 1990 *J. Vac. Sci. Technol. A* **8** 3260
[2] Matsutani T, Liu T, Mihara C, Kiuchi T, M 2004 *J Appl Phys 95** 6011
[3] Amy L. Linsebigler, Guangquan Lu, and John T. Yates, Jr. 1995 *Chem. Rev.* **95** 735
[4] Hurum D C, Agrios A G, Gray K A, Rajh T, Thurnauer M C 2003 *J. Phys. Chem.* B **107** 4545
[5] Zachariah A, Baiju KV, Shukla S, Deepa K S, James J, Warrier K G K, 2008 *J Phys. Chem. C* **112** 11345
[6] Scotti R, Arienzo M D, Testino A, Morazzoni F 2009 *Appl. Catal. B: Environ.* **88** 497
[7] Kho Y K, Iwase A, Teoh W Y, Madler L, Kudo A, Amal R 2010 *J. Phys.Chem. C* **114** 2821
[8] Mardare D, Tasca M, Delibas M, Rusu GI 2000 *Appl. Surf. Sci.* **156** 200
[9] Lazzeri M., Vittadini A, Selloni A 2002 *Phys. Rev. B* **65** 119901
[10] Singh P, Kaur D 2010 *Physica B.* **405** 1258
[11] Gua G, Li Y, Tao Y, He Z, Li J, Yin H, Li W, Zhao Y 2003 *Vacuum* **71** 487
[12] Zhang W F, He Y L, Zhang M S, Yin Z and Chen Q 2000 *J. Phys. D: Appl. Phys.* **33** 912
[13] Zeman P, Takabayashy S, 2002 *Surface and Coating Technology* **153** 93
[14] Löbl P, Huppertz M, Mergel D 1994 *Thin Solid Films* **251** 72
[15] Franco L M, Zambrano G A, Gómez M E, Camps E, Escobar-Alarcón L 2012 *Superficies y Vacío* **25(3)** 161
[16] Stafford U, Gray K, Kamat P, Varma A, *1993 Chem Phys Lett.* **55** 205
[17] Riegel G, Bolton J 1995 *J. Phys. Chem.* **99** 4215
[18] Bacsa R, Kiwi 1998 *J Appl. Catal.* B **16** 19
[19] Keiichi Tanaka, Mario F.V. Capule, Teruaki Hisanaga 1991 *Chemical Physics Letters* **187** 73
[20] Bakardjieva S, Šubrt J, Štengl V, Dianez M J, Sayagues M J 2005 *Applied Catalysis B: Environ* 58 193
[21] Seung-Min O, Takamasa I 2004 *Thin Solid Films* **457** 186
[22] Zubieta C E, German E, López Corral I, Juan A, Schulza P C, Messina PV, Dennehya M 2011 *Coll and Surf A: Physicochem. Eng. Aspects* **386** 71
[23] Bojinova A, Kralchevska R, Poulios I, Dushkin C 2007 *Mater Chem Phys* **106** 187
[24] van der Meulen T, Mattson A, Österlund L 2007 *Journal of Catalysis* **251** 131

Detailed Study of Structural and Ferroelectric Properties in $BaTiO_3/La_{2/3}Ca_{1/3}MnO_3$ Bilayers Deposited by Pulsed Laser Deposition

JE Ordoñez[1], ME Gomez[1], W Lopera[1], and P Prieto[1,2]
[1] Department of Physics, Universidad del Valle, A. A. 25360 Cali, Colombia
[2] Center of Excellence for Novel Materials - CENM, Universidad del Valle, A. A. 25360 Cali, Colombia
E-mail: john.ordonez@correounivalle.edu.co

Abstract. We have deposited ferroelectric $BaTiO_3$ (BTO), and ferromagnetic $La_{2/3}Ca_{1/3}MnO_3$ bilayers on $SrTiO_3$ substrates by pulsed laser deposition at pure oxygen atmosphere and substrate temperature of 820 °C, keeping fixed the magnetic layer thickness (t_{LCMO}) at 50 nm and varying the thickness of the ferroelectric layer (t_{BTO}) from 20 to 100 nm. X-ray diffraction analysis indicates a textured growth of both layers. The lattice parameter of the $La_{2/3}Ca_{1/3}MnO_3$ layer is slightly modified by the presence of the $BaTiO_3$ layer. Resistivity versus temperature Arrhenius plots show a small polaron thermally activated conductivity in bilayers with activation energies from 120 to 135 meV. Polarization loops at the $BaTiO_3$ measured at room temperature, using the $La_{2/3}Ca_{1/3}MnO_3$ layer as bottom electrode, show a maximum remnant polarization of 20 $\mu C/cm^2$ and minimal coercive field of 70 kV/cm. Thus, the bilayers maintain marked ferroelectric and ferromagnetic properties on individual layers.

Key words: Ferroelectric films, magnetic multilayers, magnetic properties.

1. Introduction

Multiferroic materials exhibit, simultaneously, two or more "ferroic" properties, *i.e.*, ferroelectricity (FE), ferromagnetism (FM), ferroelasticity, or ferrotoroidicity [1]. Thus, these compounds are potentially interesting for technological applications and for the study of magnetoelectric interactions. Moreover, the multiferroic property in most of the compounds exists below room temperature, limiting the practical applications. Given that both saturation magnetization and polarization are comparatively small in most intrinsic multiferroics [2,3], considerable research activity has focused on extrinsic or hybrid multiferroics based on manganites and ferroic oxides showing interesting results [4-6]. For instance, we studied the electrical and magnetic properties of ferromagnetic $La_{2/3}Ca_{1/3}MnO_3$ (LCMO) and ferroelectric $BaTiO_3$ (BTO) bilayers and their dependence on BTO thickness, maintaining constant LCMO thickness at 50 nm [7]. We observed a variation of ferromagnetic parameters like saturation magnetization and coercive field, which has been attributed to spin canting and phase separation at the BTO/LCMO interface and strain-induced magnetic disorder in the film. For a 20-nm BTO thickness, all these effects were stronger; as BTO thickness increases, the stresses at both BTO and LCMO layers is relaxed and magnetization is closer to that for an LCMO film [7]. Additionally, LCMO/BTO bilayers deposited onto Nb:$SrTiO_3$ show that the presence of a BTO film on the bilayer enhances the metal-insulator transition, T_{M-I}, probably because the strains in the LCMO layer in the bilayer are lower than those induced when the LCMO film grows directly on the substrate. When t_{BTO} increases the strains in the structure relax, allowing release of tension in the magnetic LCMO layer and improving conduction properties [8]. In this paper we report structural and ferroelectric properties of $BaTiO_3$ (t_{BTO})/$La_{2/3}Ca_{1/3}MnO_3$ (t_{LCMO})/$SrTiO_3$-substrate, when LCMO thickness is fixed (at 50 nm) and BTO thickness varies by 20, 50, and 100 nm. Transport properties were analyzed by using a small polaron model, from where activation energies were calculated. Ferroelectric properties were studied in an Au/BTO/LCMO structure at room temperature. The $BaTiO_3$ (t_{BTO})/$La_{2/3}Ca_{1/3}MnO_3$ (t_{LCMO})/$SrTiO_3$ bilayer exhibits both: ferroelectric and ferromagnetic behavior.

2. Experimental

$BaTiO_3$ (t_{BTO})/$La_{2/3}Ca_{1/3}MnO_3$ (t_{LCMO}) bilayers were deposited onto (001) single-crystal $SrTiO_3$ substrates via pulsed laser deposition (PLD) at pure O_2 atmosphere at a pressure around 10^{-3} Torr and substrate temperature of 820 °C. An excimer laser (KrF) operating at 248 nm, 10 Hz and 20-50 mJ was used. We chose three different values for the BTO thickness: 20, 50, and 100 nm, that is, t_{BTO}/t_{LCMO} is around ½, 1/1 and 2/1, respectively. The crystal structure was studied by X-ray diffraction (XRD) using a BRUKER diffractometer with Cu K_α wavelength at θ-2θ scans and rocking curves. Magnetotransport measurements were carried out in a Physical Property Measurement System (Quantum Design - PPMS–6600) in the range of 10 to 300 K. Resistance, as a function of temperature, was measured by the standard four-probe technique with current-in-plane (CIP) geometry. Magnetoresistance (MR) was examined with an applied magnetic field (H) of up to 20 kOe, perpendicular to the current direction and perpendicular to the plane. In order to compare the temperature dependence of the resistance curves, we normalized the resistance (for H = 0) at the transition temperature. For MR, we used the expression MR = R(H)/R(H=0), where R(H=0) is the resistance value at H=0 at the transition temperature. Ferroelectric hysteresis loops were carried out with an RT66B test system (Radiant Technologies) using a micromanipulator to contact the samples. Measurements were normalized with ferroelectric BTO thickness, and fatigue test show polarization cycling dependence.

3. Results

3.1. Structural and morphological properties

Figure 1 shows the out-of-plane XRD θ-2θ scans for BTO (t_{BTO})/LCMO (50 nm) bilayers grown on STO substrates, with t_{BTO} = 20, 50, and 100 nm nominal thickness. We observed the LCMO- and BTO-(001) diffraction peaks around the (001) and (002) substrate Bragg reflections, indicating a texture growth of both layers with the [001] direction perpendicular to the substrate surface. Interestingly enough, the LCMO Bragg reflex in the bilayer coincides with that for the LCMO film (dotted line) grown on STO under identical parameters. Thus, there is no effect of the BTO layer on the position of the Bragg reflex for the LCMO layer. However, the width of the (001) BTO peaks depends on its thickness being wider for the sample with t_{BTO} = 20 nm. This could indicate that for small BTO thicknesses, X-ray scans detect strains that will relax when the BTO thickness increases, and the peak is then less wide. The full width at half maximum (FWHM) of each layer in a bilayer were 0.028° and 0.09° for LCMO (50 nm) and BTO (50 nm), respectively (Fig. 1d, and 1e), which are good values for these kinds of samples and show the high quality of the bilayers obtained. The width of the rocking curve for the LCMO in a bilayer is greater than that for a 50-nm LCMO/STO film, probably because the LCMO in the bilayer is anchored at the bottom with the substrate and at the top with the BTO layer. Figure 1-f shows the BTO and LCMO lattice parameter dependence on BTO thickness. Dotted lines are plots for the *c* and *a* lattice parameters in bulk BTO; *c* substrate lattice parameter; the *c* lattice parameter in LCMO bulk and film. Plots indicate that the *c*-LCMO lattice parameter in the bilayer shifts slightly compared to that for a 50-nm LCMO/STO film, and it is independent from the thickness of the BTO layer. The *c*-BTO lattice parameter is above the BTO bulk, as expected for this thin-film material. It shows a non-monotonic decrease with increased BTO thickness, probably due to strain relaxation.

3.2. Transport properties

Figure 2a shows the dependence of ln(*R/T*) on 1/*T* (Arrhenius plot) for BTO(t_{BTO})/LCMO(50 nm) bilayers grown on STO substrate, with BTO thickness of t_{BTO} = 20 nm (open square symbol), 50 nm (open circle symbol), 100 nm (open pentagon symbol). Temperature dependence plots are the typical for the LCMO films. In the high temperature range, the curves can be described on the basis of small polaron thermally activated conductivity (SPTAC), which predicts an expression $R(T)=R_0 T\exp(E_a/k_B T)$, where E_a is the activation energy in electron volts, and k_B is the Boltzmann constant [9-11]. From the fitting of the experimental data to this expression, we calculated the activation energy. E_a decreases with increased BTO thickness (Fig. 2b). The activation energy in these bilayers is smaller than that for a 50-nm LCMO/STO film (144 meV, dotted line in plot). Changes in the activation energy, E_a, can be explained by the significant difference in the Mn–O–Mn angle in LCMO octahedra, which plays a key role for the transport mechanism in manganites [12]. We have obtained values for E_a in the range between 120 and 135 meV. It indicates that the presence of a BTO layer modifies the activation energy of the hopping conduction in the LCMO layer and may be changing the Mn-O-Mn angles in the octahedra.

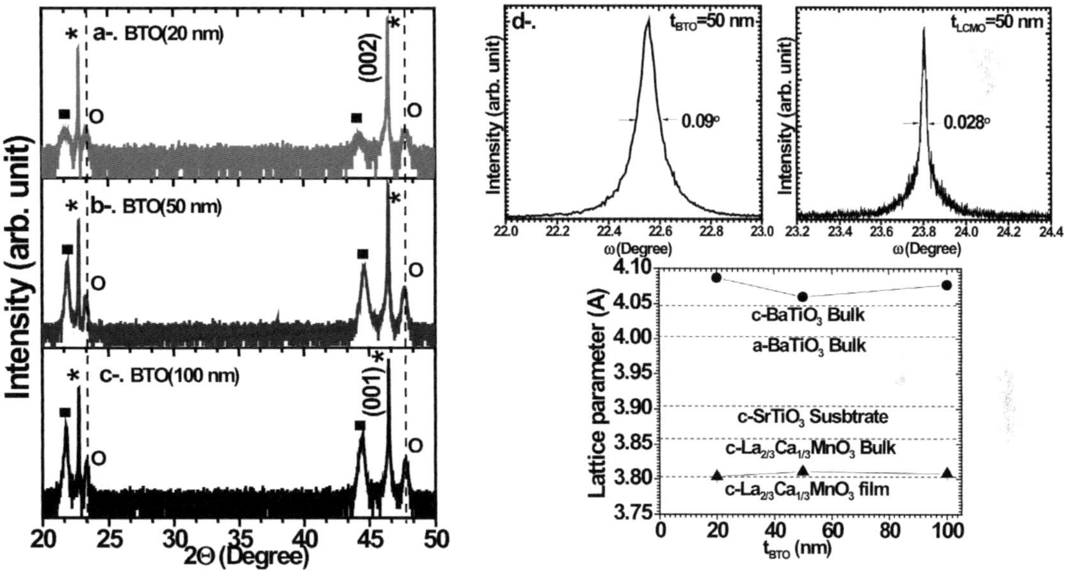

Figure 1. X-ray diffraction patterns for BTO/LCMO bilayers grown on (001) $SrTiO_3$ for a-. BTO (20 nm)/LCMO (50 nm), b-. BTO (50 nm)/LCMO (50 nm), and c-. BTO (100 nm)/LCMO (50 nm). Peaks are marked with open circles for LCMO, full squares for BTO, and asterisks for the STO substrate. d-. Rocking curve around BTO (002) peak for t_{BTO} = 50 nm with a FWHM = 0.09°, e-. Rocking curve around LCMO (002) peak for t_{LCMO} = 50 nm with a FWHM = 0.028°, f-. BTO (closed circles) and LCMO (closed triangles) lattice parameter dependence on BTO thickness. Lines are guides to the eye.

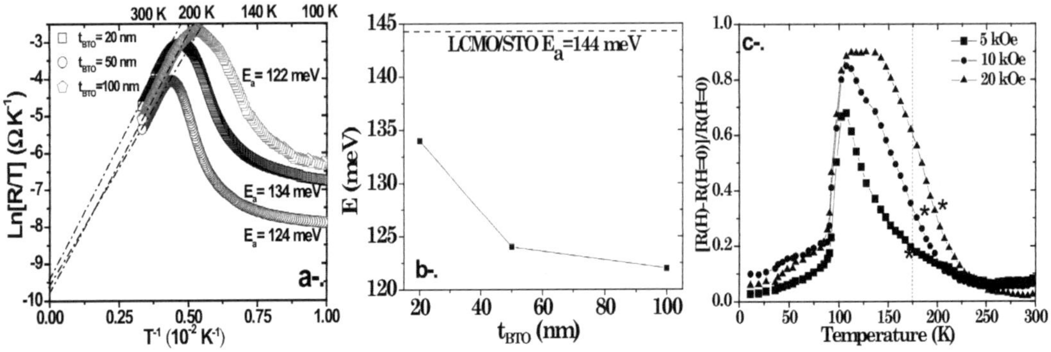

Figure 2. a-. Arrhenius plots for the BTO (t_{BTO})/LCMO (50 nm) bilayers, t_{BTO}=20 nm (open squares), t_{BTO} = 50 nm (open circles), and t_{BTO} = 100 nm (open pentagon); the dotted lines are fitting curves based on the SPTAC model. b-. Activation energy as function of BTO layer thickness (lines are guides for the eye). c-. Magnetoresistance dependence on temperature for the BTO (100 nm)/LCMO (50 nm) bilayer at applied magnetic field of 5 kOe (squares), 10 kOe (circles), and 20 kOe (triangles). Temperature for the metal insulator transition is marked with asterisks. Dotted line indicates the temperature for the metal-insulator transition (MIT) for a 50-nm LCMO film grown on (001) STO.

Figure 2c displays the temperature dependence of magnetoresistance for BTO (100 nm)/LCMO (50nm) bilayer at applied magnetic fields of 5kOe (circles), 10 kOe (squares), 20 kOe (pentagons). We have defined magnetoresistance as MR = [R(0) − R(H)]/R(0). Resistance has been measured in CIP geometry; R (H=0) and R(H) are the resistances with and without a magnetic field. In the Plots we mark with asterisks the temperature at which the MIT takes place. In all bilayers, the temperature for the MIT increases with increased applied magnetic field. This behavior is usual in optimal doped manganite [13, 14]. The temperature for the MIT is smaller than that for H=0 and that for a 50-nm LCMO film, indicating that the presence of the BTO layer affects conduction in the LCMO layer because of local structural changes at the interfaces causing competition between phases increasing the MI-width of the transition. We measured resistance in the in-plane geometry, thus, we have contributed to resistance from the BTO layer and the LCMO-BTO interface that we cannot elucidate in this kind of measurement. Magnetotransport properties are extremely sensitive to changes in double-exchange interactions, which are, in turn, strongly affected by the variations in the angles between MnO_6 octahedra induced by strain [13] at BTO/LCMO and LCMO/STO interfaces or by changes in TiO_6 octahedra crystal field.

We performed ferroelectric measurements on the Au/BTO (20-100 nm)/LCMO (50 nm) structures, using the LCMO layer as bottom electrode and Au as top electrode. Hysteresis loops were measured at room temperature, at 1-kHz frequency. Measurements were carried out under 350 kV/cm applied electric field. Figures 3a and 3b show the room-temperature ferroelectric hysteresis loops P–E for Au/BTO/LCMO structures. Both remnant polarization and coercive electric field depend on the BTO thickness. Remnant polarizations were 13, 20, and 18 $\mu C/cm^2$ and the coercive fields were 180, 103, and 70 kV/cm for t_{BTO} = 20, 50, and 100 nm, respectively, after one cycle. For the thinnest BTO layer, the remnant polarization is the smallest and coercive field is the biggest. It could indicate that, for the thin BTO layer, the electrical dipolar moment is probably diminished at the interfaces and its contribution to polarization density is smaller; whereas, the field necessary for polarization reversal increases with decreased thickness because of the strained layer that increases the pinning energy. The LCMO layer shows good performance as bottom electrode for ferroelectric measurements.

Journal of Physics: Conference Series **614** (2015) 012009 doi:10.1088/1742-6596/614/1/012009

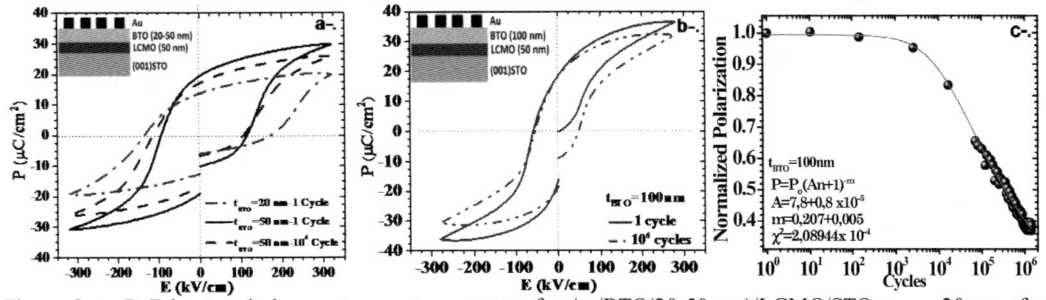

Figure 3 a-. P–E hysteresis loops at room temperature for Au/BTO(20-50 nm)/LCMO/STO. t_{BTO} = 20 nm after one cycle (dash dot), t_{BTO} = 50 nm after one cycle (solid), and t_{BTO} = 50 nm after 10^6 cycles (dash). b-. P–E hysteresis loops at room temperature for Au/BTO(100 nm)/LCMO/STO. t_{BTO} = 100 nm after one cycle (solid) and 10^6 cycles (dash dot dot). Upper inset shows the system configuration. c-. Fatigue test: normalized polarization vs. number of cycles. Solid line is the fitted curve to a $P=P_o(A_n+1)^{-m}$ theoretical model.

We performed fatigue measurements after 10^6 cycles. Figure 3c shows fatigue measurements for an Au/BTO (100 nm)/LCMO (50 nm) structure at room temperature. Normalized polarization decreases slightly with the number of cycles, but after a certain value of cycles it decreases dramatically. The behavior fit with the Yoo and Desu model [15], according to the expression $P = P_o(A_n+1)^{-m}$, for this case pilling constant $A = 7.8 \pm 0.8 \times 10^{-5}$ and decay exponent $m = 0.207 \pm 0.005$ with $\chi^2 = 2,09 \times 10^{-4}$. The effect basically consists in that for low frequencies, the oxygen vacancies will start to migrate and the trapped charges will gradually respond to the applied voltage. The rearrangement of the oxygen vacancies will affect the ferroelectric domains, impeding their switching and leading to a considerable decrease in the polarization value. The rearrangement will end after a certain number of cycles and then a recovery process might start. Recovery can occur if the amount of the charged defects acting as pinning centers for the ferroelectric domains is reduced [16,17].

4. Conclusions

We successfully produced a multiferroic hybrid system based on ferroelectric/ferromagnetic bilayers of oxide materials by using BTO as ferroelectric, and LCMO as ferromagnetic layer. The position of the Bragg reflections for LCMO films in bilayers are not modified for the presence of BTO, but the width of the (00l) BTO peaks depends on their thickness, being wider for small thicknesses, which could be related to the presence of strains that relax when BTO thickness increases. The presence of BTO layers modifies the activation energy of the hopping conduction in the LCMO layers. This can be carried out by changing the Mn-O-Mn angles in the octahedra. The metal insulator transition in LCMO layer is affected by the BTO layer because local structural changes at the interfaces can produce competition among phases, increasing the MI-width of the transition. Ferroelectric properties of bilayers show a dependence with BTO layer thickness, where for the thinnest BTO layer the remnant polarization is the smallest and coercive field is the biggest, which could indicate that for thin BTO layers, the electrical dipolar moment is probably diminished at the interfaces and its contribution to polarization density is smaller, whereas the field necessary for the polarization reversal increases with decreased thickness because of the strained layer that increases the pinning energy.

5. Acknowledgments

This work was financially supported by *"Instituto de Nanociencia de Aragón"*, Zaragoza, Spain, where the films were grown and partially characterized; the Center of Excellence for Novel Materials (CENM), COLCIENCIAS research projects: "Study of structural and physical properties in oxide materials and interfaces as potential sensor and control systems"; No 110656933104 contract No.2013-0002 CI 7917-CC 10510 COLCIENCIAS-UNIVALLE and "Study of interface phenomena in heterostructures based on complex oxides" CI 7978 UNIVALLE.

6. References

[1] Martin L, Chu Y-H, and Ramesh R 2010 *Mat. Sci. Eng R*, **68**, 89.
[2] Chu Y-H, Martin L, Holcomb M., Gajek M, Han S-J, He Q, Balke N, Yang C-H, Lee D, Hu W, Zhan Q, Yang P, Fraile-Rodríguez A, Scholl A, Wang S and Ramesh R (2008), Nat. Mater. vol 7, pp 478 – 482.
[3] Ramesh R and Spaldin N A (2007), “Multiferroics: progress and prospects in thin films”, Nat. Mater. Vol. 6, pp. 21-29.
[4] Li T, Zhang F, Fang H, Li K, Yu F (2013), Journal of Alloys and Compounds 560, pp. 167–170.
[5] Burton J D, Tsymbal E Y (2009), Phys. Rev. B 80, 174406.
[6] Ziese M (2008), J. Appl. Phys. Vol. 104, 063908/1-9.
[7] Ordoñez J E, Gómez M E, Lopera W, Prieto P (2013), IEEE Transactions on Magnetics, VOL. 49, NO. 8, pp. 4586-4589.
[8] Ordoñez J E, Gómez M E, Lopera W, Prieto P (2014), Journal of Physics: Conference Series, Volume 480, Issue 1, pp. 1-4.
[9] Navasery M, Halim S A, Dehzangi A, Soltani N, Bahmanrokh G, Erfani M, Kamalianfar A, Pan K Y, Chang S C, Chen S K, Lim K P, Awang Kechik M M, Applied Physics A, Online First, doi: 10.1007/s00339-014-8295-5.
[10] El Helali S, Daoudi K, Fouzri A, Oumezzine M, OueslatiM and Tsuchiya T (2012), Appl Phys A, vol. 108, pp. 379–384.
[11] Prokhorov V G, Komashko V A, Kaminsky G G, Svetchnikov V L, Zandbergen H W, Lee Y P, Park J S and Kim K W (2003), Low Temp. Phys. Vol. 29, pp. 117-122.
[12] Amaral V S, Lourenço A A C S, Araújo J P, Pereira A M, Sousa J B (2000), J. Appl. Phys., Vol. 87, No. 9, 1.
[13] Dagotto E, Hott T and Moreo A 2001 *Phys. Rep.*, **344**, 1.
[14] Li X, Li K, Du J, Xu X, Fang J, Wang Z, Cao X, Zhu J, Zhang Y (1998), Science in China Series A: Mathematics, Vol. 41, Issue 3 , pp 308-312.
[15]Yoo I K and Desu S B (1992), phys. Stat. sol. (a), Vol. 133, Issue 2, pp. 565-573.
[16] Scott J F and Dawber M (2000), Appl. Phys. Lett. 76, 3801-3803.
[17] Pintilie L, Vrejoiu I, Hesse D, and Alexe M (2006), Appl. Phys. Lett. 88.

Institute of Physics
Dirac House, Temple Back
Bristol BS1 6BE UK

CURRAN ASSOCIATES INC
proceedings.com